Quantum Vision: A Beginner's Guide to the Subatomic World

Om Arora

RIGI PUBLICATION

Quantum Vision: A Beginner's Guide to the Subatomic World

BY

Om Arora

Copyright© Om Arora 2023

Originally published in India

ISBN: 978-93-95773-45-4

Published by RIGI PUBLICATION

777, Street no.9, Krishna Nagar

Khanna-141401 (Punjab), India

Website: www.rigipublication.com

Email: info@rigipublication.com

Phone: +91-9357710014, +91-9465468291

To my esteemed teachers, whose wisdom guided my path,

To my beloved Mother and Father, whose love and support fueled my journey,

And to the timeless wisdom of Indian philosophy, which ignited the flame within,

This book is dedicated with profound gratitude and reverence.

PREFACE

Welcome to "Quantum Visions: A Beginner's Guide to The Subatomic World." This book is an invitation to begin on a journey through the awe-inspiring world of physics, specifically focusing on the fascinating realm of quantum mechanics.

Physics is the fundamental science that explores the mysteries of the natural world, from the microscopic building blocks of matter to the vast expanse of the cosmos. It is a discipline that encompasses both the grandest cosmic phenomena and the tiniest particles, revealing the principles that govern our reality.

In this introductory guide, we will delve into the realm of quantum physics—a realm that challenges our intuitions and takes us beyond the everyday experiences of classical physics. Quantum mechanics introduces us to a world where particles can exist in multiple states simultaneously, where observation influences the behavior of objects, and where uncertainty is woven into the fabric of reality.

But fear not! While quantum physics may seem daunting at first, this book aims to present the concepts and principles in a manner that is accessible and engaging, without sacrificing the essence of this intricate and beautiful subject. As we journey together, we will explore the foundations of quantum mechanics, examine famous experiments that shaped our understanding, and uncover the intriguing applications of quantum physics in our daily lives.

We will begin on a into the heart of the atom, investigating the wave-particle duality and the uncertainty principle. We will encounter mind-bending phenomena such as quantum superposition and entanglement, challenging our perceptions of reality. From there, we will expand our horizons to explore the fields of particle physics, cosmology, and the quest for a unified theory.

Throughout this book, my aim is not only to convey knowledge but also to instill a sense of wonder and curiosity about the world around us. I want to ignite the spark of inquiry within you, encouraging you to ask questions, seek answers, and embrace the beauty of scientific exploration.

Whether you are a student embarking on your first encounter with physics or an enthusiast eager to deepen your understanding, "Quantum Visions" promises to be a companion that will guide you through the intricate tapestry of quantum mechanics. Together, we will unravel the enigmatic nature of the universe, exploring its mysteries.

So, let us embark on this journey together, armed with curiosity and an open mind. Prepare to be amazed as we venture into the depths of quantum realms, guided by our shared quantum visions.

Om Arora

ACKNOWLEDGEMENTS

I would like to express my deepest gratitude to all those who have contributed to the creation of this book, "Quantum Visions."

First and foremost, I would like to extend my heartfelt gratitude to the following teachers and mentors of mine who are like God in my life:

Mrs. Ritu Bahri, Mrs. Kathy, Mrs. Bharti Bhateja, Mrs. Pallavi, Mrs. Ambika, Mrs. Ayushi and my inspiration Mr. Manish Purohit.

I am grateful to the scientific community for their groundbreaking research and discoveries, which have paved the way for our understanding of the quantum world. Their tireless efforts and dedication to advancing knowledge have been instrumental in shaping the content of this book.

I extend my heartfelt appreciation to the authors, researchers, and scientists whose works have been referenced throughout this book. Their valuable insights, theories, and experimental findings have served as the foundation for the topics discussed, enriching the content and providing a comprehensive understanding of the subject matter.

I would like to thank the editors, proofreaders, and reviewers who have contributed their expertise, time, and meticulous attention to detail to ensure the accuracy and clarity of the content. Their invaluable feedback and suggestions have greatly improved the quality of this book.

I would also like to express my gratitude to Adarsh Public School and its Principal, Mrs. Pooja Malhotra, for providing a nurturing environment and fostering a culture of academic excellence.

I would also like to express my appreciation to my family and friends for their unwavering support and understanding. Their love, encouragement, and belief in me have been a constant source of motivation during this endeavor.

Lastly, I am deeply grateful to the readers of this book. It is my sincere hope that "Quantum Visions" serves as a valuable resource, igniting curiosity and fostering a deeper appreciation for the wonders of quantum mechanics and its implications for our understanding of the universe.

Thank you all for your invaluable contributions and for being a part of this journey.

With sincere gratitude,

Om Arora

CONTENTS

Chapter 1: Introduction to Quantum Physics   **11**

1.1 The Birth of Quantum Mechanics   11

1.2 Wave-Particle Duality   15

1.3 The Uncertainty Principle   19

1.4 Quantum States and Probability   24

Chapter 2: Exploring Quantum Phenomena   **34**

2.1 The Double-Slit Experiment   34

2.2 The Observer Effect   40

2.3 Quantum Superposition   44

2.4 Quantum Entanglement   51

Chapter 3: Quantum Applications in Everyday Life   **61**

3.1 Quantum Computing   61

3.2 Quantum Cryptography   70

3.3 Quantum Communication   81

3.4 Quantum Sensors and Metrology   85

Chapter 4: The Quantum World of Particles   **91**

4.1 The Standard Model of Particle Physics   91

4.2 Elementary Particles and Forces   99

4.3 Particle Accelerators and Colliders   106

4.4 The Higgs Boson and the Origins of Mass   112

Chapter 5: Seeking the Theory of Everything   **121**

5.1 Supersymmetry and Super partners   121

5.2 String Theory and the Multiverse   131

5.3 Quantum Gravity and Black Holes   140

5.4 The Holographic Principle   151

Conclusion: Embracing Quantum Visions   **162**

Further Reading: Recommended Resources for Further Exploration   **165**

Glossary: Key Terms and Concepts   **168**

Chapter 1

Introduction to Quantum Physics

1.1 Birth of Quantum Mechanics

Introduction

Welcome to the fascinating world of quantum mechanics, which has revolutionised our understanding of the universe. In this section, we'll go back in time to investigate the creation of quantum physics, a scientific revolution that forever changed our perspective of reality. Prepare to be fascinated as we explore the foundational discoveries and mind-bending paradoxes that lead to the birth of this amazing branch of physics.

A Crisis in Classical Physics

Classical physics, which had reigned supreme for centuries, was in crisis at the turn of the twentieth century. Phenomena such as blackbody radiation, the photoelectric effect, and atomic behaviour defy explanation within the framework of classical theories. Scientists were confronted with challenges that necessitated new approaches and insights.

The problem of blackbody radiation, for example, amazed scientists who tried to comprehend how the strength of radiation emitted by a hot object depended on its temperature. Classical theories projected that when the wavelength of the radiation approached zero, the intensity would increase without bound, a process known as the "ultraviolet catastrophe." However, experimental observations contradicted these assumptions, necessitating a new understanding.

Planck's Quantum Hypothesis

In 1900, Max Planck presented a new theory that would usher in the quantum revolution. Planck discovered the concept of quantized energy by analysing blackbody radiation, in which energy is emitted and absorbed in discrete packets known as "quanta." This ground-breaking notion established the groundwork for a new age of physics.

Planck's quantum theory proposed that energy is not continuous but exists in discrete amounts. He proposed that the energy of each quantum is directly proportional to the frequency of the radiation, and he provided a fundamental constant, now known as Planck's constant, to establish the link. This new knowledge contradicted the long-held notion that energy could take any arbitrary value and opened the door to a completely new understanding of the physical world.

Planck's revolutionary hypothesis not only provided a theoretical explanation for the puzzling behavior of blackbody radiation but also set the stage for further breakthroughs in the realm of quantum mechanics.

Einstein's Light Quanta

Building on Planck's quantum theory, Einstein achieved a significant breakthrough in 1905 by explaining the phenomenon of the photoelectric effect. The photoelectric effect is the emission of electrons from a substance when it is irradiated by light.

According to classical wave theory, the energy delivered to electrons should grow linearly with light intensity. However, experimental results revealed that the maximal kinetic energy of released electrons was determined by the frequency of the incident light rather than its intensity.

Einstein proposed in his remarkable paper that light is made up of discrete particles called "photons," each of which carries a specific amount of energy. The frequency of light determines the energy of a single photon.

This realisation challenged the established wave theory of light, laying the framework for the theory of wave-particle duality.

Einstein's groundbreaking work on the photoelectric effect earned him the Nobel Prize in Physics in 1921, solidifying quantum theory's foundation.

The Bohr Model of the Atom

Niels Bohr developed a revolutionary model of the atom that integrated quantum principles in 1913. The model accurately explained the discrete nature of atomic spectra and introduced the concept of electron energy levels, where electrons exist in quantized orbits around the atomic nucleus.

According to classical physics, electrons orbit the nucleus in the same way that planets orbit the sun, eventually emitting energy and spiralling into the nucleus. However, this concept failed to explain atom's stability and the absence of continuous radiation. This issue was addressed by Bohr's model.

Bohr proposed that electrons have discrete energy levels and that transitions between these levels correlate to the emission or absorption of photons. The quantized values of angular momentum determine the energy of each level.

The Bohr model advanced our understanding of atomic structure and lay the framework for future quantum advances.

Wave-Particle Duality

As quantum theories evolved, scientists battled with the puzzling nature of particles and waves. In 1924, Louis de Broglie proposed his famous wave-particle duality hypothesis, which postulated that particles such as electrons might have both wave-like and particle-like qualities.

De Broglie's idea was inspired by the already established wave-particle duality of light. De Broglie postulated that particles, like photons, may behave as both waves and particles. This concept challenged the conventional conception of particles as localised entities and opened up a new line of study into the quantum nature of reality.

The concept of wave-particle duality represented a significant shift in our understanding of the underlying nature of matter. It demonstrated that particle behaviour cannot always be represented only by classical notions, necessitating the development of a new framework to comprehend their nature.

Heisenberg's Uncertainty Principle

Werner Heisenberg introduced the Uncertainty Principle in 1927, an idea that would forever change our understanding of the microscopic world. According to Heisenberg's principle, there are limitations in our capacity to properly measure both a particle's position and momentum at the same time.

In quantum physics, the wave-like nature of particles leads to the Uncertainty Principle. It states that the more accurately we try to measure one observable, the less accurately we know the other. This fundamental limit emphasises the probabilistic aspect of quantum physics and challenges determinism.

Heisenberg's Uncertainty Principle paved the way for a new understanding of the uncertainty and indeterminacy present in the microscopic world. It explained that quantum mechanics would deal with probabilities and statistical descriptions, rather than precise deterministic measurements.

Schrödinger's Wave Equation

In the same year, Erwin Schrödinger developed his famous wave equation, which defined the behaviour of quantum systems in terms of wavefunctions. The wave equation established the mathematical framework for quantum physics by allowing the calculation of particle probabilities and energy levels.

Schrödinger's equation is a partial differential equation that determines the time development of the wavefunction. It describes the wave-like behaviour of particles and how their wavefunctions evolve over time. The equation gives a mathematical framework for calculating the probabilities of detecting a particle in a given state or location.

Schrödinger's wave equation revolutionized the domain of quantum mechanics by providing a powerful tool for describing the behaviour of particles in terms of probabilities. It enabled physicists to make predictions about the behaviour of quantum systems and paving the way for further advancements in quantum theory.

As we reach the end of this exploration into the birth of quantum mechanics, we have touched upon the foundational concepts that marked the birth of quantum mechanics. From Planck's quantum hypothesis to Einstein's light quanta, and from Bohr's atomic model to Heisenberg's Uncertainty Principle, we have glimpsed the remarkable ideas that revolutionized our understanding of the microscopic world.

1.2 Wave-Particle Duality

Introduction

Now let us take a look at the world of wave-particle duality, where the boundaries between matter and energy, particles and waves, blur into a path of scientific research challenging our knowledge of the microscopic world. Here, in this section, we start an epic journey of discovery, exploring the mysteries and mind-bending phenomenon that define the principle of wave-particle duality. Prepare to be immersed in a boundless pool of knowledge, as we travel through the complexities of this profound concept.

The Dual Nature of Light

Our journey begins with an investigation of light, a phenomena that has fascinated humans for ages. Light was traditionally thought of as a wave—a radiant object travelling across space, defined by its wavelength, frequency, and propagation qualities. This wave model successfully explained a variety of optical phenomena, ranging from light bending through lenses to the generation of colourful rainbows.

However, as scientists explored much deeper into the nature of light, the story took an unexpected turn. The mystery began to open-up when Thomas Young did his famous double-slit experiment in the early nineteenth century. Young discovered an extraordinary interference pattern—an undeniable indication of wave-like behaviour—by allowing light to pass through two closely spaced slits and analysing the ensuing pattern on a screen. This discovery gave compelling proof that light had wave-like properties, allowing for interference and diffraction.

The story of light's duality, however, did not end there. The photoelectric effect, methodically investigated by Albert Einstein and others in the early twentieth century, broke the prevalent idea that light was only made up of waves. When light interacted with specific materials, it exhibited particle-like behaviour, as proved by the photoelectric effect. This discovery gave rise to the concept of photons—discrete packets of energy—in which light behaved as if it were made up of distinct particles, each carrying a certain quantity of energy. This discovery was a paradigm shift, challenging conventional understanding of light and establishing the concept of wave-particle duality.

Particle-Wave Duality: Pioneers and Experiments

As we progress deeper into the quantum realm, we come across the fascinating concept of particle-wave duality—a concept that goes beyond light and covers the basic essence of matter itself. In 1924, Louis de Broglie published his revolutionary idea, inspired by the parallelism between light and matter: all particles exhibit both particle-like and wave-like properties.

De Broglie's ground-breaking discovery proposed that particles, despite their apparent solidity and localization, show wave-like behaviour. A particle's wavelength, which is inversely proportional to its momentum, became a defining feature of its wave-like character. This breakthrough forced scientists to investigate particle wave-like behaviour, challenging the belief that particles were only entities confined to a fixed location.

A variety of experiments were carried out to confirm the wave-like character of particles. Notably, electron diffraction experiment done in 1927 by Clinton Davisson and Lester Germer gave solid evidence for electron wave-like behaviour. They discovered that electrons diffracted, when they were accelerated through a potential difference and aimed towards a crystal, producing interference patterns that were very similar to those seen in Young's double-slit experiment with light. This experiment demonstrated that particles, in this case electrons, had wave-like properties, confirming the theory of wave-particle duality.

Similar tests were carried out with protons, neutrons, and even larger things such as atoms and molecules. Wave-like phenomena, including as diffraction, interference, and wave-particle interactions, were consistently seen in these studies. The scientific community was presented with the indisputable fact that particles have both particle and wave natures at the quantum level, a realisation that changed our understanding of the fundamental building blocks of the cosmos.

The Uncertainty Principle: Limits of Knowledge

As we advance through wave-particle duality, we come into one of quantum physics' most profound principles: the Uncertainty Principle, famously proposed by Werner Heisenberg in 1927. The Uncertainty Principle sets fundamental constraints on the precision with which certain complementary properties of a particle can be measured at the same time.

The Uncertainty Principle presents itself as a basic constraint in our ability to precisely define both the position and momentum of a particle in the context of wave-particle duality. The more precisely we seek to measure a particle's position, the less exactly we can determine its momentum, and vice versa. This uncertainty comes as a result of particles' wave-like nature, in which their properties become essentially probabilistic rather than deterministic.

The Uncertainty Principle fundamentally challenges our classical intuitions, showing the inherent limits of knowledge and measurement at the quantum level. It challenges us to confront the idea that accurate predictions and measurements of particles are essentially restricted by the probabilistic character of quantum physics. While this may appear to be annoying at first, it opens up a universe of exciting possibilities, paving the way for a greater understanding of the delicate fabric of the quantum world.

As we approach the end of this exploration of the complex interplay between waves and particles that defines the fabric of our reality. The astonishing realisation that matter and energy, at their most fundamental level, have both wave-like and particle-like properties arouses in us a great sense of surprise and interest.

The beauty of scientific research and the vast frontiers of knowledge that lie ahead begs us to question the fundamental essence of our existence, as wave-particle duality calls us to do. It inspires us to broaden our perspectives, question our assumptions, and delve deeper into the mysteries that rule the quantum world.

1.3 The Uncertainty Principle

Introduction

The Uncertainty Principle as discussed briefly in previous section is a fundamental notion in quantum mechanics that transformed our view of the physical world. Here, in this section, we start on a profound examination of uncertainty, exploring its conceptual landscape and exploring into its consequences for the quantum realm. We explore into the complexities of uncertainty, igniting the spark of curiosity and cultivating a profound respect for the complicated nature of quantum mechanics, guided by our commitment to scientific inquiry.

Heisenberg's Matrix Mechanics: A Framework for Uncertainty

The Uncertainty Principle is better understood with the help of Heisenberg's matrix mechanics. We will go through a brief overview of it and try to understand its implications, which introduces us with the concept of non-commutativity and replaces classical observables with matrices.

The concept of mathematical matrices, which represents quantum states and observables, is central to Heisenberg's Matrix Mechanics. Unlike in classical mechanics, where physical quantities are defined by simple numbers, quantum mechanics introduces the concept of operators acting on mathematical vectors. These operators, represented by matrices, allow us to perform calculations and make predictions about the behaviour of quantum systems.

Matrix Mechanics replaces the classical concept of precise, deterministic values with a probabilistic framework guided by mathematical matrices. These matrices, which encode observables and quantum states, provide a language for describing the behaviour and interactions of quantum systems.

Uncertainty

As we have discussed Particles have both wave-like and particle-like properties, according to wave-particle duality.

So, when we describe a quantum entity as a wave, we can determine its momentum with reasonable accuracy but when we want to precisely define its position, we must regard it as a localised particle.

The contradiction between these two methods of measurement gives rise to the Uncertainty Principle. To accurately detect the position of a quantum particle, we must restrict it to a small region of space. By doing so, however, we effectively construct a localised wave packet made of several waves of varying momenta. As a result, precisely determining the particle's position results in a loss of knowledge about its momentum.

Consider the example of a wave packet made of several component waves, to better understand this phenomenon. Each component wave represents a different momentum, which contributes to the overall wave packet. When we try to determine the particle's position, we see a localised concentration of the wave packet at a single point. However, this localization leads in a superposition of several momentum components, making it hard to determine the particle's exact momentum.

Complementarity: The Interplay of Wave and Particle Aspects

The concept of complementarity is crucial for understanding the dual nature of particles. Complementarity refers to the relationship between the wave and particle portions of quantum entities, offering an fascinating concept that contradicts classical intuitions. This concept, first proposed by Niels Bohr, emphasises the fundamental principle that certain aspects of a quantum system cannot be observed with precision at the same time.

The wave-particle duality that characterises quantum phenomena' behaviour demonstrates that particles exhibit both wave-like and particle-like properties. When seen as waves, particles exhibit features like as diffraction and interference, which are similar to the behaviour of waves in classical physics. When viewed as particles, they have distinct and localised properties, similar to classical particles with well-defined positions.

Because of complementarity, it is fundamentally impossible to measure both the wave and particle parts of a quantum phenomenon at the same time. This is due to the underlying mathematical nature of quantum mechanics, in which measurements affect the system under observation. When we choose to measure wave-like features, such as interference patterns, we cannot collect exact information on the particle's position at the same time. If we concentrate on determining the particle's position, we lose information regarding its wave-like properties, such as wavelength or frequency.

Experimental Realizations of Uncertainty

As mentioned earlier the double-slit experiment is one of the most well-known experiments for demonstrating wave-particle duality and the significance of uncertainty. In which a beam of particles, such as photons, is directed towards a barrier with two small slits in this experiment. An interference pattern appears when the particles flow through the slits and onto the screen, indicating wave-like behaviour. This phenomena suggests that particles have wave-like properties and can interfere with one another, resulting in zones of constructive and destructive interference. When detectors are used to determine which slit a particle travels through, the interference pattern disappears and particle-like behaviour takes over. This experiment highlights the mutual exclusivity of concurrently monitoring both the wave and particle characteristics, emphasising the underlying uncertainty in determining the route and position of the particle.

The Stern-Gerlach experiment is another ground-breaking experiment that shows uncertainty. This experiment investigates the quantization of spin angular momentum and the difficulties in measuring its components. A beam of particles, typically silver atoms, is steered through an inhomogeneous magnetic field in the setup. As the particles move through the field, their spin orientation gets aligned with either the up or down direction of the magnetic field. Surprisingly, the particles are observed to divide into distinct beams, revealing quantized spin states, rather than a continuum of possible spin orientations. This experiment reveals that measuring the spin in one direction inevitably destroys the information about the spin in the opposite direction, highlighting the uncertainty inherent in calculating several spin components at the same time.

Furthermore, the famous Einstein-Podolsky-Rosen (EPR) experiment and subsequent Bell's theorem have been crucial in understanding the difficulties of locality and non-locality in quantum mechanics. Two particles are entangled in these experiments, and their properties become fundamentally linked. Measurements on one particle have an immediate effect on the properties of the other particle, regardless of their distance. This non-local correlation challenges our classical notions of locality and emphasises the intrinsic uncertainty and entanglement of quantum systems.

Experiments such as quantum entanglement, quantum teleportation, and quantum cryptography also highlight the significance of uncertainty in practical applications. These discoveries have had far-reaching technological consequences, paving the path for quantum computing, secure communication, and high-precision measurements.

Technological Implications: Uncertainty and Quantum Technologies

While the Uncertainty Principle challenges our understanding of the underlying basis of reality, it also throws us in a universe of technological possibilities. Here we explore into the technological consequences of uncertainty, notably in the domain of quantum technology. We look at two essential areas: quantum computing and quantum cryptography, where uncertainty plays an important role in improving processing capacity and maintaining secure communication channels.

Quantum Computing:

Quantum computing is one of the most promising applications of uncertainty. Traditional computers use bits, which can be in one of two states: 0 or 1. Quantum computers, on the other hand, use the principles of uncertainty and superposition to generate quantum bits, or qubits, which can exist in several states at the same time. This superposition of states allows quantum computers to conduct tasks in parallel, increasing their processing capability in comparison to traditional computers.

Uncertainty serves as the foundation for quantum algorithms and quantum gates, which are the building blocks of quantum computers. Quantum computers may handle complicated tasks more efficiently than their classical counterparts by utilizing the uncertainty of quantum systems. They are particularly good at tasks like factoring big numbers, simulating quantum systems, and optimising complex algorithms.

While quantum computing is still in its infancy, it has enormous promise to alter industries such as encryption, material research, drug development, and optimisation. The advancement of quantum algorithms and error-correction techniques continues to push the boundaries of uncertainty, paving the road for realistic quantum computers capable of tackling real-world problems.

Quantum Cryptography:

Another important technological implication of uncertainty is found in quantum cryptography. Traditional cryptography systems rely on mathematical techniques that take into account the computing difficulties of specific tasks. However, as powerful quantum computers and advances in mathematical algorithms become available, these systems may become vulnerable.

Quantum cryptography, on the other hand, employs principles of uncertainty to ensure secure communication routes. Quantum key distribution (QKD) is one such technology that uses the uncertainty in quantum measurements to establish a shared secret key between two parties. QKD ensures the security of the key exchange by encoding information in quantum states, as any interception or measurement attempt would produce noticeable disturbances.

Because quantum measurements are fundamentally unpredictable, quantum cryptography is very safe against eavesdropping or hacking efforts. It provides verifiable security based on quantum physics' laws, allowing for safe communication even in the presence of formidable attackers.

The development of viable quantum cryptography systems is paving the way for secure communication routes, particularly in financial, government, and defence sectors. Quantum key distribution mechanisms are being used to protect sensitive information while also maintaining data confidentiality and integrity.

1.4 Quantum States and Probability

Introduction

Here we are going to look into the fascinating world of quantum states and probability in this section. We examine how mathematical concepts known as wavefunctions explain quantum systems and how these wavefunctions encode the probabilities of alternative outcomes when measurements are done. Understanding quantum states and probability is essential for appreciating quantum mechanics' probabilistic character and the uncertainty inherent in quantum occurrences.

Quantum States

Understanding the concept of quantum states is important in understanding the behaviour and characteristics of quantum systems in the framework of quantum mechanics. Quantum states are the fundamental descriptions of quantum entities, encompassing their unique properties and guiding their behaviour.

The Wavefunction:

To express quantum states in quantum mechanics, we use a mathematical tool called the wavefunction. The wavefunction captures the complicated relationship of quantum phenomena' wave-like and particle-like properties. We get insight into the features and behaviour of each quantum system by assigning it a unique wavefunction.

The wavefunction is a complex-valued function that is dependent on coordinates of the system. It is denoted by the symbol Ψ (psi) and represents the probability of locating the system in a specific condition. The probability density of finding the system in a specific state upon measurement is given by the square of the wavefunction, $|\Psi|^2$.

Superposition:

Superposition is a thrilling phenomenon in quantum mechanics. It enables quantum systems to exist in several states at the same time. To put it another way, a quantum system can be in two or more states at the same time, with each state contributing to the overall wavefunction.

The linearity of quantum physics gives birth to superposition. The wavefunctions of the different states combine to generate a composite wavefunction, preserving the system's coherence. This unique characteristic allows quantum systems to exhibit behaviours that defy conventional logic, opening up new possibilities for quantum information processing and quantum computing.

Complex Numbers:

The utilisation of complex numbers to express the amplitudes of quantum states is an important characteristic of quantum physics. Complex numbers are a powerful mathematical tool for capturing both magnitude and phase, allowing us to understand quantum phenomena' interference and wave-like characteristics.

The wavefunction's complex numbers provide useful information about the probability and relative phases of distinct states within the superposition. We can control and utilise the quantum features of systems by controlling these complex amplitudes, leading to significant applications in quantum technology.

Measurement and Probability in Quantum Mechanics

Measurement is important to scientific investigation, allowing us to learn about the physical world. Measurement is critical in retrieving information from quantum systems in the world of quantum physics. However, quantum measurements differ greatly from classical measurements. Now, we will look at the concept of probability in quantum mechanics and how it relates to the wavefunction. The Born Rule, which provides the mathematical link between the wavefunction and the probability of measurement outcomes, will be investigated. We will investigate how quantum probabilities materialise in experiments and the significance of statistical interpretations in quantum measurements.

In contrast to classical physics, where measurements produce accurate results, quantum measurements are inherently probabilistic. The wave-like characteristics of quantum systems and the probabilistic interpretation of the wavefunction contribute to this essential element. Quantum measurements give statistical information about potential outcomes, reflecting the inherent uncertainty in quantum events.

Max Born's Born Rule provides the mathematical relationship between the wavefunction and the probabilities of measurement results. The Born Rule states that the probability of receiving a specific measurement outcome is proportional to the squared magnitude of the corresponding component of the wavefunction.

If we have a wavefunction Ψ, the probability for obtaining a measurement result that corresponds to a specific observable is determined by $|\Psi|^2$. The probability density distribution across the possible measurement results is represented by this quantity. We can calculate the probability of obtaining different measurement findings by computing the squared magnitudes of the wavefunction's components.

Now let's consider the double-slit experiment to see how quantum probabilities emerge in experiments. Particles such as electrons or photons are delivered through two slits in this experiment, resulting in an interference pattern on a screen behind the slits. The distribution of particle hits on the screen corresponds to the wavefunction's probability.

When the particles pass through the slits, they behave like waves, creating an interference pattern. However, when measured, each particle behaves as a localised entity, displaying on the screen as a point-like particle. The wavefunction's probability distribution precisely forecasts the probability of each particle striking a specific point on the screen.

The nature of reality and the role of observation are fascinating tasks brought up by quantum probability. To explain the probabilistic nature of quantum measurements, various statistical interpretations have been offered. One of the most common interpretations holds that "The act of measurement collapses the wavefunction, resulting in a single definite measurement output".

Other interpretations, such as the many-worlds interpretation and the pilot-wave theory, provide different perspectives on how probabilities are interpreted in quantum mechanics. These interpretations investigate the possibility of many worlds or hidden variables governing the behaviour of quantum systems.

The Collapse of Wavefunction

The collapse of the wavefunction is a key idea in quantum physics that sits at the heart of measurement. When a quantum system is measured, the wavefunction that defines its state experiences a quick and unanticipated transformation, "collapsing" into one of the potential measurement outcomes. Here, we will look at the process of wavefunction collapse, as well as the role of observers and measurement equipment. We will look at wavefunction collapse interpretations, such as the Copenhagen interpretation and the many-worlds interpretation, as well as the ongoing arguments about the nature of the collapse process.

The role of observers and measurement apparatus in the collapse of the wavefunction is critical in quantum physics. The act of measuring introduces interaction between the quantum system and the measuring apparatus, causing the wavefunction to collapse.

When a measurement is made by an observer, the quantum system becomes entangled with the measurement device, resulting in an entangled state. The measurement device amplifies microscopic quantum effects to macroscopic sizes, allowing us to view and record measurement results.

Niels Bohr and his colleagues proposed the Copenhagen interpretation, which is one of the most widely accepted interpretations of wavefunction collapse. The wavefunction, according to this view, describes the probabilities of measurement results, and measurement causes the wavefunction to collapse into one of the potential outcomes.

The collapse of the wavefunction is viewed as an inherent aspect of the measuring process, linked to the act of observation in the Copenhagen interpretation. The interaction of the observer with the quantum system causes the collapse, resulting in a definite measurement output.

Hugh Everett's many-worlds interpretation offers an alternative viewpoint on wavefunction collapse. According to this interpretation, the wavefunction does not collapse. Instead, the wavefunction splits into several branches, each of which corresponds to a particular measurement outcome.

According to the many-worlds interpretation, every conceivable consequence of a measurement exists in its own branch of reality, resulting in an endless number of parallel universes. Each universe takes its own course, retaining the wavefunction's coherence over all possibilities.

Quantum Entanglement And Its Implications

Quantum entanglement is a fascinating phenomenon that is very important to quantum mechanics. It explains an unique correlation that can emerge between quantum systems, in which their states become inseparably connected regardless of distance. Here, we will look into entanglement and its extraordinary qualities, as well as the implications for our understanding of quantum physics. The renowned EPR paradox and Bell's theorem will be discussed, which provide light on the non-local nature of entanglement and challenge our preconceived notions of locality and causation. Furthermore, we will investigate real-world entanglement experiments and applications, such as quantum teleportation and quantum communication, to demonstrate the exciting possibilities that result from harnessing this fascinating quantum phenomena.

Entanglement happens when the states of multiple quantum systems become correlated to the point where the behaviour of one system is inherently connected to the behaviour of the other, regardless of physical distance. This association maintains even when the systems are separated by great distances, challenging our traditional concepts of local interactions.

Quantum entanglement is a result of the superposition principle and quantum mechanics' mathematical formalism. When two or more particles get entangled, their combined state cannot be defined simply as the sum of their individual states. The entangled system, on the other hand, exists as a whole, with its state tightly linked across its constituent particles.

Einstein, Podolsky, and Rosen developed the EPR paradox as a thought experiment to test the completeness of quantum physics. It emphasised the entanglement's paradoxical aspect, as well as its implications for localization and hidden variables.

Two particles are formed in an entangled state, with their attributes interwoven, in the EPR thought experiment. When one particle is measured, establishing its state instantly collapses the wavefunction of the other particle, regardless of the distance between them. This observation points to a non-local connection in which information appears to flow faster than the speed of light, challenging traditional concepts of causation.

The EPR paradox sparked much debate and research into the nature of entanglement, eventually leading to the formulation of Bell's theorem.

Bell's theorem, discovered by physicist John Bell, provided a mechanism to experimentally verify quantum mechanics predictions against local reality, a concept that asserts the existence of hidden variables affecting quantum system behaviour.

Scientists have proved that the predictions of quantum physics and entanglement violate the inequalities derived from local realism through innovative experiments known as Bell tests. These experimental results firmly demonstrate that the correlations observed in entangled systems cannot be explained by classical hidden variables, reinforcing quantum entanglement's non-local nature.

Entanglement is not only a fascinating aspect of quantum mechanics; it also has significant consequences for technology and communication. Entanglement has been used by researchers to create ground-breaking applications such as quantum teleportation and quantum communication.

Quantum teleportation is the transmission of a particle's quantum state to another distant entangled particle. Although it does not involve physical transportation, this phenomenon allows for the faithful duplication of quantum information over space, opening up possibilities for safe and efficient communication.

Quantum communication uses the unique qualities of entanglement to build safe channels for information transmission. Quantum communication protocols provide a better level of security than traditional encryption methods by utilising the non-local correlations of entangled particles, making them invaluable for sensitive applications.

Quantum States in practice: Measurement Techniques and Quantum Information Processing

Now, we turn our attention to the practical features of quantum states and their manipulation. We investigate the measurement techniques used to retrieve information from quantum systems, as well as the state preparation process. This investigation provides insights into the practical applications and future prospects of quantum states.

Measurement is a crucial feature of quantum mechanics that allows us to extract information about quantum systems' properties. Here, we will look at different measurement techniques that scientists employ to examine and characterise quantum states. Understanding these techniques can provide us with important insights into the behaviour and features of quantum systems.

In quantum mechanics, measurements are made using observables, which are mathematical operators that represent physical quantities. Observables can correlate to numerous quantum system attributes such as location, momentum, energy, or spin. Each observable is connected with eigenvalues and eigenvectors, which are critical for comprehending measurement results.

The measuring process can change the quantum state when done on a quantum system. The collapse of the wavefunction is the name given to this occurrence. The measuring procedure leads the wavefunction to "collapse" into one of the eigenstates associated with the observable under consideration. This collapse gives us a precise measuring result.

The deliberate manipulation and engineering of quantum systems into specified quantum states is a critical aspect of quantum mechanics. Now, we will look at the methods and strategies that scientists use to shape quantum states for experimental purposes. Understanding state preparation strategies can help us understand active control and manipulation of quantum systems.

Optical pumping, which uses laser light to manipulate the quantum states of atoms or ions, is one way of state preparation. Optical pumping can be used to selectively align atomic spins, resulting in specific quantum states. This technique is very beneficial for atomic clocks, quantum magnetometers, and other applications that require the production of well-defined quantum states.

Coherent control is a powerful tool for manipulating quantum states precisely. It entails the use of customised electromagnetic fields, such as laser pulses, to induce desired states in quantum systems. Scientists can use coherent control to modify quantum states, create superpositions, and even execute quantum gates for quantum computer applications. This technology has transformed the world of quantum control and opens up new possibilities for quantum information processing.

Quantum state tomography is a technique for reconstructing an entire description of a quantum state. It involves taking a series of measurements on a group of identically prepared quantum systems and analysing the collected data to determine the underlying quantum state.

Conclusion

As we conclude this chapter we have discussed the origin of quantum mechanics, and the fundamental principles which govern it. This provides us with a basic overview of the subject allowing us to understand the mysterious characteristics of the quantum realm.

Chapter 2

Exploring Quantum Phenomena

2.1 The Double-Slit Experiment

Introduction

As in the previous chapter we have seen some of the very basics and foundations of Quantum Mechanics, This section here will explore into one of quantum physics' most exciting experiments, the double-slit experiment. This landmark experiment gives information on the wave-particle duality of quantum phenomena and calls into question our fundamental concept of reality. We examine the deep complications of the double-slit experiment and its role in shaping our understanding of quantum mechanics through a series of fascinating discoveries and research.

The Setup: Two Slits and One Mystery

The double-slit experiment fascinates scientists and beginners alike by exposing the mysterious nature of quantum mechanics. The experiment is based on a simple setup consisting of a barrier with two small slots and a screen placed behind it. To carry out the experiment, a source emits particles, such as photons, that are aimed against the barrier.

As these particles flow between the two slits, an amazing event occurs: an interference pattern appears on the screen behind the barrier. This pattern is similar to the classic ripples generated when two waves overlap and interact. Scientists are puzzled by the fact that this interference pattern persists even when particles are passed through the slits one at a time over a long length of time.

The emergence of an interference pattern proves that the particles behave in a wave-like manner. It implies that each particle that passes through the slits behaves like a wave, spreading out and interfering with itself. The resulting pattern on the screen shows alternating bright and dark bars, which represent constructive and destructive interference, respectively.

The fascinating feature of the double-slit experiment is that when particles are transmitted separately, one would expect them to behave as discrete entities, resulting in two distinct groups on the screen corresponding to the positions of the slits. Instead, the particles interfere with one another, as if they have wave-like qualities.

This strange behaviour contradicts our conventional beliefs that particles should follow predictable trajectories and create different effects. The double-slit experiment, illustrates the complicated nature of particles in the world of quantum physics, blurring the line between waves and particles.

To investigate deeper into the riddle, scientists have carried out variations of the double-slit experiment with particles of varying sizes and energy, such as electrons or even huge molecules like buckyballs. Surprisingly, the interference patterns persist across particle kinds, supporting the idea that wave-like behaviour is a fundamental feature of the quantum environment.

Scientists have presented different theories and interpretations in their attempts to understand the complexities of the double-slit experiment. These range from wave-particle duality, which implies that particles have both wave and particle properties, to more complex interpretations including parallel universes and hidden variables. The double-slit experiment has set the way for major advances in the realm of quantum physics by serving as a launching pad for studying the fundamental nature of reality.

The Interference Pattern

The appearance of an interference pattern on the screen is one of the most stunning characteristics of the double-slit experiment. This pattern demonstrates the wave-like behaviour of quantum phenomena and calls into question our classical notions of particle motion.

When particles, such as electrons or photons, pass through the barrier's two slits, they behave like waves, spreading out and interfering with one another. The superposition of distinct wavefunctions associated with the particles causes this interference phenomena.

Waves from the two slits overlap and interact as they propagate outward. Waves align and reinforce one another in specific areas, resulting in constructive interference. This constructive interference causes places on the screen where the wave amplitudes add together, resulting in bright spots or areas of increased intensity.

In other places, the waves from the two slits may cancel out, resulting in destructive interference. When the peaks of one wave align with the troughs of another, the amplitudes subtract, resulting in dark spots or areas of reduced intensity on the screen.

The interference pattern that appears on the screen is a direct result of the particles' wave-like behaviour. It has a sequence of light and dark fringes that change in intensity and spacing. The wavelength of the particles and the distance between the slits dictate the spacing between the fringes, which follows a well-defined mathematical relationship.

Visualisation approaches and mathematical tools, such as wave equations and superposition principles, are used to better understand the complexities of interference. These techniques enable us to predict and analyse interference patterns generated by various experimental settings.

The presence of an interference pattern in the double-slit experiment is a clear indication of quantum phenomena' wave-like nature. It presents convincing proof that particles have wave qualities and experience interference, contradicting our traditional notion of discrete particles travelling along deterministic paths.

Even when particles are sent through the slits separately, the discovery of an interference pattern shows the fantastic phenomena of wave-particle duality. It implies that, while quantum entities exhibit particle-like behaviour when measured, they exist in a superposition of several states, similar to waves, until their interactions are noticed.

Particle Detection

While the interference pattern in the double-slit experiment demonstrates the wave-like behaviour of quantum entities, introducing particle detectors changes the observed phenomena dramatically. Here, we look at the component of the experiment in which detectors are used to determine the direction that the particles take.

When detectors are placed at or behind the slits, they serve as tools for determining which path the particles take. As the wavefunction associated with the particles collapses, the act of measurement causes a substantial change in their behaviour.

Prior to measurement, the particles exist in a state of superposition, exhibiting wave-like behaviour and interference. When a detector is applied, however, the wavefunction collapses, causing the particles to act as discrete particles with fixed positions. This collapse is caused by the interaction of the particles with the measurement apparatus, which causes the particles to be localised to specific routes.

The addition of particle detection has a significant impact on the observed interference pattern on the screen. When detectors are present, the particles can no longer interact with each other in a wave-like manner. They instead behave as different entities, with distinct trajectories defined by the measurement outcome.

The previously detected interference fringes starts to decrease as the particles lose their capacity to interfere. The pattern on the screen changes from a series of bright and dark fringes to a collection of clear, localised dots that correspond to the places of the particles identified.

This shift from wave-like to particle-like behaviour emphasises the delicate nature of quantum measurements. The process of measuring destroys the particles' underlying superposition and entanglement, revealing their distinct character. It highlights a fundamental concept of quantum mechanics: measuring disrupts the system being measured.

In quantum physics, the elimination of interference fringes in the presence of particle detection indicates the complementarity of wave and particle characteristics. We sacrifice the ability to see interference when we attempt to gather information about the path taken by the particles. This trade-off between information and interference is central to wave-particle duality and the inherent uncertainty in quantum measurements.

Delayed Choice and Quantum Eraser

The double-slit experiment not only violates the classical notion, but it also presents mind-bending occurrences that call our understanding of causality into question. Now we explore into the notions of delayed choice and quantum Eraser, which show the experiment's non-local correlations and retroactive consequences.

The delayed choice experiment adds a twist to the design by giving the user the option of detecting or erasing the path information of the particles after they pass through the slits. Surprisingly, this choice has an effect on the observed interference pattern on the screen.

Consider the following scenario: detectors are put at the slits to determine the trajectories of the particles, interrupting the interference pattern. However, after the particles have travelled through the slits and reached the screen, the decision to detect or not detect them can be made.

Surprisingly, the interference pattern disappears when the decision to detect is taken, and the particles act like distinct particles with localised positions. If the choice is made not to detect, the interference pattern reappears, as if the particles had never contacted any detecting apparatus.

The experiment's delayed decision feature calls into question our preconceived views of causality. It implies that current decisions can have an impact on events that have already occurred in the past. The experiment's outcome appears to be influenced by a retroactive influence, in which the measurement choice made after the particles have passed through the slits influences their behaviour and the observed interference pattern.

The quantum Eraser experiment extends the delayed choice notion by investigating the relationship between particle path detection and the ability to observe interference. By inserting additional quantum systems known as "entanglement-based quantum erasers," the interference pattern can be restored even after the path information has been measured.

Entangled particles are formed alongside particles travelling through the slits in the quantum eraser arrangement. These entangled particles carry complementary information about the original particles' journeys. It is feasible to re-establish the interference pattern on the screen by manipulating the entangled particles and selectively removing or retrieving the path information.

Non-local correlations and the interrelated character of quantum systems are revealed by the delayed choice and quantum eraser experiments. They propose that information regarding a particle's route, even if it appears to be measured or deleted in the present, can have retroactive consequences on the observed interference pattern. These occurrences call into question our classical notion of cause and consequence, pointing at quantum mechanics' complicated and elusive nature.

2.2 The Observer Effect

Introduction

Now we shall explore the observer effect and its quantum mechanics consequences. We investigate how the act of observation or measurement changes the behaviour of quantum systems, challenging our understanding of the observer-observed relationship. We set out to solve the mystery surrounding the observer effect and its significance in the quantum world.

Observer Effect: Shaping Reality through Measurement

The observer effect is a fascinating phenomenon at the heart of quantum mechanics. It demonstrates the significant relationship between the observer and the observed system, implying that the act of observation or measurement can affect the properties and behaviour of a quantum system. The implications of the observer effect challenge our conventional beliefs and force us to think again about the nature of reality at the microscopic level.

The observer effect dates back to the early twentieth century, when quantum mechanics was being developed as a new theory to explain particle behaviour at the atomic and subatomic scales. Pioneers such as Werner Heisenberg and Niels Bohr shaped our understanding of the observer effect and its importance in quantum measurement.

The Uncertainty Principle, proposed by Heisenberg in 1927, was an invaluable contribution that shed light on the limitations of simultaneous measurements of certain pairs of physical characteristics, such as position and momentum, with arbitrary precision. According to Heisenberg's principle, the more precisely we attempt to determine one property, the more uncertain the measurement of the other property becomes. This inherent uncertainty in measuring emphasised the close relationship between the act of observation and the disturbance induced to the observed system.

Niels Bohr expanded on the observer effect concept in his Copenhagen interpretation of quantum physics. The act of measuring, according to Bohr, causes the wavefunction, which reflects the quantum state of a system, to collapse into one of the potential measurement results. This collapse indicates that the system's attributes become definite only when they are observed, resulting in the emergence of classical reality from the world of possibilities.

The observer effect calls into question our traditional notion of objective reality existing independently of observation.

Quantum Decoherence

The phenomenon of decoherence in quantum mechanics sheds light on the delicate interaction between a quantum system and its surrounding environment. Decoherence is crucial for understanding the observer effect and the transition from quantum to classical behaviour. Now, we will look at the idea of quantum decoherence and how it relates to the observer effect.

When a quantum system interacts with its surroundings, which might be made up of particles, fields, or other external elements, decoherence occurs. This interaction causes entanglement between the system and its surroundings, resulting in loss of coherence and quantum state superposition. The decoherence process brings a progressive transition from the quantum realm, which is characterised by wave-like behaviour and superposition, to the classical world, where particles have specific attributes and behave conventionally.

The quantum system becomes entangled with a huge number of degrees of freedom as a result of its interaction with the environment, such as particle locations or field fluctuations. The impacts of tiny quantum uncertainty are amplified by this entanglement, resulting in macroscopic irreversibility and the emergence of classical reality. As a result, the delicate quantum interference patterns and superpositions that occur in isolated systems are suppressed, giving the appearance of classical behaviour.

Quantum decoherence explains why macroscopic objects, which are made up of many particles, behave classically despite being fundamentally governed by quantum mechanics. When a quantum system becomes entangled with its surroundings, its wavefunction interacts with an infinite number of other wavefunctions, causing interference effects to cancel out and resulting in a loss of coherence. As a result, the system appears to be classically behaving, with well-defined attributes such as position and momentum.

The importance of decoherence in the observer effect cannot be overstated. When a quantum system is measured, the measurement apparatus becomes entangled with the system and its surroundings. The system rapidly loses coherence due to the interaction with the environment, resulting in the collapse of the wavefunction and the appearance of a clear measurement conclusion. This explains why macroscopic objects appear to have definite features when seen, because their quantum nature is swamped by the effects of decoherence.

Understanding and regulating decoherence is critical for a variety of quantum technology applications, including quantum computing and quantum information processing. Researchers are working to find strategies to reduce or lessen the consequences of decoherence, allowing for the preservation of sensitive quantum states and increasing the promise of quantum technology.

Quantum Zeno Effect

The quantum Zeno effect is a fascinating phenomena that comes from the interaction of measurement and quantum system dynamics. Now, we look at how frequent measurements or continuous observation might stymie a quantum system's natural evolution or change, challenging our intuition regarding the observer effect.

The quantum Zeno effect is named after the ancient Greek philosopher Zeno of Elea, who presented a series of motion-related paradoxes. The quantum Zeno effect refers to the inhibition or "freezing" of a quantum system's evolution when it is frequently observed or monitored in the context of quantum mechanics.

The evolution of a quantum system is guided by a unitary transformation, which defines how the system's state changes over time, according to quantum mechanics principles. When a system is continuously measured, however, the act of observation impacts its unitary evolution and causes it to remain in its initial state.

Consider the wavefunction collapse that occurs during measurement to understand the quantum Zeno phenomenon. Each measurement effectively resets the system's evolution by projecting it onto one of its conceivable states. If measurements are taken frequently enough, the system has little opportunity to move away from its starting condition, resulting in change suppression.

In quantum physics, this counterintuitive result raises important concerns regarding the nature of observation and the significance of measurement. We seem to inhibit the evolution of a quantum system by continuously observing it, despite not making any direct changes to its state.

The quantum Zeno effect has been demonstrated experimentally utilising many quantum systems, including atoms, photons, and superconducting qubits. These experiments involve repeatedly measuring and observing the system's behaviour. The observed freezing or inhibition of the system's evolution serves as empirical evidence for the quantum Zeno effect's existence.

In quantum mechanics, the quantum Zeno effect has also been related to the concept of time. It challenges the nature of time and the role of measurement in defining its passing. The continuous observation of a quantum system calls into question the concept of a continuous and unbroken passage of time because it appears to "halt" the system's evolution.

It is crucial to stress, however, that the quantum Zeno effect does not imply a total prohibition of change in a quantum system. Rather, it is the result of the unique conditions imposed by frequent measurements. The system can resume its normal evolution if the measurement frequency is reduced or the measuring process is adjusted.

The quantum Zeno effect has implications in a variety of domains, including quantum computing and quantum information processing. It emphasises the complex relationship between measurement, observation, and quantum system dynamics. Researchers can acquire insights into the fundamental nature of measurement and the limits it imposes on the evolution of quantum systems by researching and comprehending the quantum Zeno phenomenon.

2.3 Quantum Superposition

Introduction

The concept of quantum superposition, which defies traditional intuition and offers a radical shift in our understanding of reality, is at the heart of quantum mechanics. Here we shall explore deeper into the interesting phenomena of quantum superposition, in which quantum systems can exist in numerous states at the same time. We investigate superposition's mathematical formalism, the experimental evidence for its existence, and implications for our knowledge of the quantum world.

Mathematical Formalism

Understanding the behaviour of quantum systems requires a formal definition of quantum superposition. The concept of a quantum state, which represents the qualities and characteristics of a quantum system, is central to this explanation. A quantum state is often represented in quantum mechanics by a mathematical construct known as a wavefunction.

And as mentioned earlier a wavefunction, indicated by the notation Ψ (psi), is a complex-valued function that encodes quantum system information. It includes all important information about the system's location, momentum, energy, and other observable quantities. The wavefunction is defined throughout the full space in which the system is capable of existing.

The ability of a quantum state to exist in a superposition of several states is one of its fundamental features. This indicates that a quantum system can exist in multiple states with varying probabilities at the same time. This is expressed mathematically as a linear combination of basis states. The base states represent the system's distinct and well-defined states. In the case of a spin-1/2 particle, for example, the basic states may be spin-up (denoted as $|\uparrow\rangle$) and spin-down (denoted as $|\downarrow\rangle$).

A general quantum state $|\psi\rangle$ is a linear combination of the following base states:

$$|\psi\rangle = \alpha\,|\uparrow\rangle + \beta\,|\downarrow\rangle$$

where α and β are complex numbers referred to as probability amplitudes. The probability of detecting the system in a given condition is proportional to the square of its probability amplitude. In this case, $|\alpha|^2$ represents the probability of measuring the spin-up state, and $|\beta|^2$ represents the probability of measuring the spin-down state. It is crucial to remember that the sum of all conceivable measurement outcomes must equal 1.

The complex nature of probability amplitudes characterises quantum mechanics. It enables the interaction of probability amplitudes, resulting in phenomena such as constructive and destructive interference. This interference produces the distinctive patterns seen in quantum experiments, such as the interference pattern shown in the double-slit experiment.

Consider the following example to better understand superposition and probability amplitudes. Assume we have a quantum system that can exist in two states, $|0\rangle$ and $|1\rangle$. The system's state can be represented as follows:

$$|\psi\rangle = \alpha|0\rangle + \beta|1\rangle$$

The probability of achieving the states $|0\rangle$ and $|1\rangle$. are given by $|\alpha|^2$ and $|\beta|^2$, respectively, if the system is measured. Because the total probability must be one, $|\alpha|^2 + |\beta|^2 = 1$.

The coefficients α and β are complex numbers, and their phases are important in defining the interference effects seen in quantum systems. When the phases of α and β are such that they move parallel to each other, constructive interference occurs, increasing the probability of certain events. When the phases oppose each other, destructive interference develops, resulting in a decreased probability of detecting certain events.

Quantum superposition allows for the coexistence of several states and the simultaneous observation of various possibilities. It is a fundamental idea in quantum physics that explains many quantum events and technologies. We acquire a greater understanding of the complicated nature of quantum systems and their behaviour by using the mathematical framework of superposition.

Experimental Evidence

Experimental evidence is an important pillar in proving the validity of quantum superposition. Over the years, numerous tests have been carried out, providing solid evidence for the coexistence of multiple quantum states. Let's look at some important experiments that demonstrate superposition and wave-particle duality.

As discussed before, Stern-Gerlach experiment was one of the first to demonstrate quantum superposition. A beam of particles, such as silver atoms, is passed across a magnetic field gradient formed by a magnet in this experiment. According to classical physics, the magnetic field would deflect the particles equally. However, the testing results indicated an unexpected result. The beam divided into distinct directions, suggesting that the particles had intrinsic angular momentum. As also mentioned before, another key experiment that demonstrates the wave-like behaviour of particles and the interference caused by quantum superposition is the double-slit experiment. In which a beam of particles, such as electrons or photons, is directed towards a barrier having two microscopic slits in this experiment. A screen behind the barrier records the arrival of the particles. According to classical physics, the particles would flow through one of the slits or the other, resulting in two unique patterns on the screen. However, an interference pattern identical to that created by waves is detected. This interference pattern depicts the superposition of states, in which particles travel through both slits simultaneously and interfere with one other, resulting in constructive and destructive interference patches on the screen.

Experiments utilising quantum systems like atoms, ions, and photons trapped in quantum systems like optical lattices or ion traps have also offered convincing evidence for superposition. Researchers have proved the ability to create and control superposition states by changing the states of these systems using lasers and other control techniques, resulting in visible quantum interference effects.

This data not only proves the presence of quantum superposition, but also shows the wave-particle duality inherent in quantum systems. The observed interference patterns, as well as the ability to modify quantum particle states, provide compelling evidence for the coexistence and interference of numerous quantum states.

Furthermore, technological breakthroughs and practical implementations of quantum technologies, such as quantum computing and quantum communication, provide additional evidence for the existence and utility of quantum superposition. To obtain computing and communicational benefits over conventional systems, these technologies rely on the capacity to manage and utilise the superposition of quantum states.

Entangled Superposition

Entangled superposition is a fascinating element of quantum physics that occurs when several particles' quantum states become entangled, resulting in a highly connected and inseparable system. Now we explore the notion of entangled superposition and its far-reaching consequences in quantum research and technology.

Entanglement happens when the quantum states of two or more particles become entangled to the point where one particle's state cannot be defined independently of the others. This results in a highly entangled superposition state in which the system's overall state cannot be divided into individual states of the constituent particles. Instead, the system exists in a coherent superposition of all potential particle-state combinations.

The formation of non-local correlations is one of the most exciting elements of entangled superposition. This means that no matter how far apart they are, measuring the state of one entangled particle instantly affects the state of the other. This non-locality stands in stark contrast to classical physics, which assumes that information and effect spread at a limited speed determined by the speed of light.

Entangled superposition has profound implications in many areas of quantum research and industry. Entanglement is a crucial resource in quantum information processing for accomplishing computational tasks that are beyond the capabilities of classical systems. Quantum techniques, such as Shor's algorithm for factoring huge numbers and Grover's algorithm for exploring unsorted databases, use entangled superposition to produce exponential speedups over classical algorithms. We shall discuss about these algorithms in detail in the upcoming chapters.

Entanglement is also important to quantum communication methods like quantum teleportation and quantum key distribution. The quantum state of one particle can be transferred to another distant particle using the entangled superposition between them in quantum teleportation. Quantum key distribution, on the other hand, uses the non-local correlations of entangled particles to construct secure communication channels because any attempt to intercept would disrupt the entanglement, leaving signs of interception.

Superposition and Measurement

In the realm of quantum physics, quantum superposition and measurement are inseparable from one another. Now we explore the complex link that exists between these two essential components of quantum theory.

A quantum system in superposition exists in a state that is a composite of numerous potential states. When the system is measured, the superposition state collapses into one of the possible outcomes, producing a definite measurement result. This collapse distinguishes quantum measurement from conventional measurement procedures.

The act of measuring forces the system to "choose" a specific state, and the probability associated with different measurement outcomes is determined by the amplitudes of the superposition states' wavefunctions. The Born Rule states that the probability of receiving a specific measurement outcome is given by the squared magnitude of the corresponding amplitude. The probabilistic character of quantum measurement reflects the quantum level's inherent uncertainty and indeterminacy.

It is vital to note that the measurement apparatus and the interaction between the system and the measuring device have an impact on the measurement process. This interaction has the potential to disrupt the fragile quantum superposition, resulting in measurement-induced decoherence. The interaction of the system with its surroundings causes the loss of quantum coherence and the emergence of classical behaviour, which is referred to as decoherence.

Measurement-induced decoherence can cause the superposition state to seem to "collapse" into a classical mixture, in which the system operates as a probabilistic collection of classical states rather than a coherent superposition. The macroscopic appearance of classical reality in our everyday experience is frequently due to the rapid development of decoherence.

The interaction between quantum superposition and measurement has been thoroughly examined and experimentally validated. The well-known double-slit experiment, for example, shows how the act of measurement (sensing of the particle's position) collapses the superposition state and results in the appearance of particle-like behaviour, with unique interference or diffraction patterns.

Understanding measurement and its relationship to quantum superposition is critical for understanding quantum phenomena and developing practical applications in quantum information processing and quantum technology. Quantum algorithms and protocols use superposition and measurement features to achieve computing speedup and secure communication, respectively.

Applications and Implications

The concept of quantum superposition has profound implications and can be applied in a variety of domains. Now, we'll look at some of these applications and talk about the broader implications of quantum superposition.

Quantum computing is one of the most exciting uses of quantum superposition. Quantum computers make use of quantum bits, also known as qubits, which can exist in superposition states. Unlike classical bits, which can only be in one of two states, qubits can represent both 0 and 1 at the same time due to superposition. This superposition capability allows quantum computers carry out computations on a massively parallel scale. Quantum algorithms, such as Shor's algorithm for factoring huge numbers and Grover's algorithm for exploring unsorted databases, as mentioned earlier, use superposition to solve complicated problems faster than traditional computers. The potential influence of quantum computing on cryptography, optimisation, and quantum system modelling is huge with the potential to revolutionise these domains.

Superposition also enables the phenomena of quantum parallelism, in which a quantum system can investigate numerous possibilities at the same time. This has implications for optimisation issues, where quantum algorithms can search for the best solution by investigating numerous potential solutions at the same time. Furthermore, quantum simulation uses superposition to describe and investigate complicated quantum systems that are difficult to simulate with classical computers. These applications hold the possibility of major improvements in a variety of scientific and technical disciplines by utilising the capabilities of superposition and quantum parallelism.

2.4 Quantum Entanglement

Introduction

Quantum entanglement is another fascinating phenomenon at the heart of quantum mechanics. Now, we will look at the concept of quantum entanglement and how it might help us comprehend the nature of reality. We investigate the fascinating non-local correlations that emerge when two or more particles get entangled, challenging classical notions of locality and separability.

Entanglement

Entanglement is a unique event in quantum mechanics that defies conventional understanding. It represents a situation in which the properties of two or more particles become intrinsically linked to the point that one particle's state cannot be explained independently of the others. Entanglement, or interconnectedness, is a fundamental element of quantum theory that has significant implications for the nature of reality.

Entanglement is caused by the superposition of quantum states. Because of the principle of superposition, particles in quantum mechanics can exist in several states at the same time. Individual states of two or more particles that interact and become entangled can no longer be represented separately. Instead, their aggregate state is made up of all potential combinations of individual states.

The entangled state of a system is represented mathematically by a tensor product that combines the states of the constituent particles. For example, if particle A can be in state $|a\rangle$ and particle B can be in state $|b\rangle$, the system's entangled state is written as $|\Psi\rangle = |a\rangle \otimes |b\rangle$, where $\otimes$ indicates the tensor product.

It is critical to understand that entanglement is a non-local phenomenon. This means that entangled particles can exhibit correlations that are not explained by local interactions or classical theories. Regardless of the distance between them, the entangled particles appear to have an instantaneous link. This non-local aspect of entanglement has been experimentally demonstrated, and it poses fundamental challenges to our traditional understanding of space, time, and causation.

The effect of entanglement on measurements is one of the most intriguing features of the phenomenon. When an observable on one entangled particle is measured, it instantly changes the state of the other entangled particle, regardless of the distance between them. Numerous investigations have proved this phenomenon, known as quantum non-locality, which is a fundamental property of entanglement.

Numerous theoretical and experimental studies have been triggered by the entangled state and its non-local nature. Scientists have investigated several forms of entangled states, such as maximally entangled states such as the Bell states, and devised techniques for producing and managing entanglement. Entanglement research has also resulted in substantial advances in quantum information processing, with applications in quantum computing, quantum communication, and quantum cryptography.

EPR Paradox

Albert Einstein, Boris Podolsky, and Nathan Rosen postulated the Einstein-Podolsky-Rosen (EPR) paradox in 1935. It was created to put certain aspects of quantum mechanics to the test and shed light on the non-local nature of entanglement. The EPR paradox shaped our knowledge of entanglement and helped to the advancement of quantum mechanics as a field.

The thought experiment begins with an entangled pair of particles known as the EPR pair. The properties of these particles, according to quantum theory, are associated even when they are separated by large distances. When these features are measured, the EPR paradox develops.

Einstein, Podolsky, and Rosen claimed that measuring the properties of one particle would offer instant knowledge about the properties of the other particle, regardless of their distance. This seems to imply faster-than-light transmission, which appears to defy Einstein's theory of relativity, which states that information cannot be transmitted faster than the speed of light.

The paradox arises from the disagreement between quantum mechanics' projected non-local correlations and the concept of locality, which states that information may only be conveyed at or below the speed of light. The EPR thought experiment tested scientists' ability to reconcile these seemingly contradictory beliefs.

To address the EPR problem, physicists created Bell's theorem and, later, Bell's inequalities. These mathematical formulations enabled experimental testing of quantum mechanics predictions and determination of whether non-local correlations of entanglement were consistent with classical explanations.

Several experiments have been done to test Bell's inequalities, and the findings have consistently supported quantum mechanics predictions. These investigations demonstrated the truth of entanglement and its consequences for our understanding of physical reality by confirming non-local correlations and violating Bell's inequalities.

The EPR paradox and related experiments had a significant impact on the advancement of quantum physics. They caused the development of entanglement-based technologies such as quantum cryptography, quantum teleportation, and quantum computing by recognising entanglement as a fundamental element of nature.

Bell's Theorem

Bell's theorem is a basic conclusion in quantum physics that illuminates the non-local interactions that exist in entangled systems. It was developed in the 1960s by physicist John Bell and has since become a cornerstone of quantum theory. Bell's theorem gives a mathematical framework for checking quantum mechanics predictions against the concept of local realism.

The concept of a Bell inequality, which is a mathematical inequality resulting from particular assumptions regarding local realism, is introduced in the theorem. According to local realism, physical characteristics of objects continue independently of observation and that information cannot travel faster than the speed of light. These principles support classical physics and indicate that any correlations between distant particles must adhere to specific constraints.

However, Bell's theorem shows that these assumptions are incompatible with quantum mechanics predictions. It demonstrates that entangled particles can display correlations that beyond the bounds of classical physics. In other words, Bell's theorem demonstrates that entangled particle correlations cannot be explained by local hidden variables.

Physicists have designed experimental setups known as Bell tests to test Bell's theorem. These studies include measuring the properties of entangled particles in various ways and comparing the results to local realism predictions. If the measured correlations contradict the Bell inequality, this implies the presence of non-local correlations and supports quantum mechanics concepts.

Numerous Bell tests have been performed over the years, with constant violations of the Bell inequality. This experimental evidence demonstrates the non-local nature of entanglement and supports quantum mechanics predictions. In different systems, including photons, electrons, and ions, breaches of the Bell inequality have been detected, providing strong evidence for the presence of non-local interactions.

The importance of Bell's theorem stems from its role in ruling out local hidden variable theories, which contend that there are underlying, unseen variables that influence measurement results. According to Bell's theorem, such theories can't match the correlations observed in entangled systems. This suggests that quantum mechanics' predictions, with their non-local correlations, are more consistent with experimental observations.

Bell's theorem has significant implications for our understanding of the quantum world. It proves that entanglement is a real physical phenomenon with non-local connections between particles, not just a mathematical abstraction. The violations of the Bell inequality challenge our conventional assumptions about locality, causality, and the nature of reality itself.

Furthermore, the confirmation of non-local correlations using Bell tests has cleared the way for the development of practical applications including quantum cryptography, quantum teleportation, and quantum communication. It has opened up new possibilities for utilising entanglement in quantum technology, as well as sparked deeper research into the mysteries of quantum entanglement.

Quantum Teleportation

Quantum teleportation is a fascinating idea that facilitates the transmission of quantum states between particles that are separated by huge distances. It is based on entanglement principles and quantum mechanics characteristics. Now, we will look at quantum teleportation and its implications.

The procedure begins with a pair of entangled particles, often known as an entangled resource or an entangled pair. These particles are created in a particular quantum state that exhibits the entanglement property. The sender keeps one of the entangled particles, known as the entangled state or the Bell state, while the other particle, known as the target or the teleported state, is transmitted to the receiver.

The purpose of quantum teleportation is to move an unknown particle's quantum state, known as the input state, to the teleported state at the receiver's location. Any unique quantum state, such as the state of a qubit, can be used as the input state. Teleportation is accomplished through a set of measurements and procedures performed by both the sender and the recipient.

The sender initiates the teleportation by performing a joint measurement on the input state and the entangled state. This is referred to as a Bell measurement, and it includes measuring specific properties of the two particles. This measurement yields a classical result, which the transmitter conveys to the receiver via classical channels.

When the receiver receives the measurement result, it applies a set of quantum operations known as quantum corrective operations on the teleported state. These operations are determined by the measurement result supplied by the sender. By performing these actions, the receiver effectively teleports the quantum information by transforming the transported state into an exact replica of the input state.

The unique feature of quantum teleportation is that the quantum state is transferred without physically transferring the input state. Instead, the entangled pair serves as a resource that allows information to be transferred instantly, in a process that appears to be quicker than the speed of light. It should be noted, however, that information transfer is vulnerable to the no-cloning theorem, which asserts that it is impossible to generate a perfect copy of an unknown quantum state.

Quantum teleportation has profound implications for quantum information processing. It facilitates distributed quantum computing and quantum communication protocols in the realm of quantum computing by allowing qubits to be transferred between separate places. It is also important in quantum cryptography, where the secure transport of quantum information is critical.

Furthermore, quantum teleportation enables distant modification of quantum states, allowing for remote measurements and operations on quantum systems. This opens up possibilities for remote sensing, quantum simulations, and other applications requiring distant manipulation of quantum states.

Quantum Communication

Quantum communication is a branch of science that uses the unique qualities of entanglement to create secure and efficient information transmission. Now we will look at the fundamentals and applications of quantum communication.

The vulnerability of information to interception and hacking is one of the core issues in traditional communication. Quantum communication addresses this issue by utilising quantum mechanics concepts to construct safe channels for information transmission.

Quantum key distribution (QKD) is an important protocol in quantum communication. QKD enables two parties, typically referred to as Alice and Bob, to establish an unconditionally secure shared encryption key. The security of the key distribution mechanism is based on entanglement features.

In QKD, Alice prepares and sends a sequence of entangled particles to Bob. These particles could, for example, be photons in a certain quantum state. Because the particles are entangled, any attempt to intercept the broadcast would disrupt the entanglement, leaving traces of the interceptor's existence. Alice and Bob can then measure their individual particles and compare the data to identify any possible interception attempt.

Alice and Bob can distil a final encryption key that is known only to them and safe against any interception efforts through a process of information reconciliation and privacy amplification. This key is then usable for secure communication via traditional encryption techniques.

The challenge of conveying quantum information over vast distances is another essential aspect of quantum communication. Entanglement is a delicate process that is easily disrupted by environmental noise and transmission losses. This is a substantial barrier to long-distance quantum communication.

To address this issue, the concept of quantum repeaters was developed. Quantum repeaters are devices that can expand the range of quantum communication across great distances by preserving and distributing entanglement. They accomplish this by segmenting the transmission and applying entanglement switching procedures.

Entangled pairs of particles are formed and distributed via intermediate stations in a quantum repeater system. Entanglement swapping between neighbouring pairs is performed by these stations, thereby extending entanglement between distant pairs. Entanglement can be retained and transferred over larger distances by repeating this procedure iteratively.

Quantum repeaters, which minimise the effects of noise and transmission losses, show enormous potential for enabling effective long-distance quantum communication. However, research towards efficient and dependable quantum repeater technology is ongoing.

Entanglement and Quantum Information Processing

Entanglement is crucial in quantum information processing, providing distinct advantages over classical computation. Now, we will look at how entanglement is used for computing tasks and how it is used in diverse sectors.

The quantum gate, which is an operation that manipulates the state of a quantum system, is a critical component in quantum information processing. When qubits are entangled, these gates can operate on many qubits at the same time, resulting in parallel calculations and exponential processing capacity.

Entanglement allows quantum algorithms to outperform their classical counterparts in specific tasks. Quantum algorithms use entanglement properties to solve problems more effectively, such as Shor's algorithm and Grover's algorithm, which are quite popular and celebrated than others. These algorithms use entanglement to examine numerous options at the same time, giving them a major advantage over traditional algorithms.

Entanglement is also important in quantum simulation, which uses quantum systems to simulate and explore complicated processes. Researchers can build highly entangled states that represent the behaviour of the system being emulated by entangling qubits. This enables more realistic description of quantum dynamics, allowing for research into quantum chemistry, materials science, and other topics where classical simulation methods face substantial challenges.

Another area where entanglement has uses is quantum optimisation. Researchers can use entanglement to search for optimal solutions more effectively by encoding optimisation problems into the entangled states of qubits. Quantum algorithms, such as the Quantum Approximate Optimisation Algorithm (QAOA), use entanglement to explore the solution space in ways that traditional optimisation techniques cannot.

Entanglement is also important in quantum error correction, which is an important part of quantum information processing. Researchers can safeguard quantum states from noise and decoherence by encoding information in entangled states and using error-correcting codes. The entanglement of qubits enables the identification and correction of errors without interfering with the encoded data.

Chapter 3

Quantum Applications in Everyday Life

3.1 Quantum Computing

Introduction

So as of now we have discussed almost all basic concepts and principles of Quantum World and you must have got some clues about their implications. So now in this chapter, we will take a look at the interesting topic of quantum technologies, which uses quantum physics principles to revolutionise computation. We investigate the fundamental concepts underlying quantum computing, its potential advantages over traditional computing, and the difficulties associated with harnessing its power and we shall cover all the engineering and technological aspect of the Quantum Mechanics.

Principles of Quantum Computing

Quantum computing is based on quantum mechanics principles, which regulate the behaviour of particles at the quantum level. The concept of qubits, the quantum counterpart of classical bits, is central to quantum computing. Qubits differ from classical bits in numerous important aspects, the most significant of which is their ability to live in superposition and get entangled with other qubits.

Superposition:

One of the distinguishing characteristics of qubits is their ability to exist in many states at the same time, known as superposition. Unlike traditional bits, which can only represent a 0 or a 1, qubits can be in a linear mixture of both 0 and 1. A qubit in superposition is expressed mathematically as $|\psi\rangle = \alpha|0\rangle + \beta|1\rangle$, where α and β are complex probability amplitudes that define the probabilities of measuring the qubit in the 0 or 1 state. This superposition enables quantum computers to process information in parallel and investigate numerous possibilities at the same time.

Entanglement:

Another essential concept in quantum computing is entanglement. When qubits get entangled, their states become coupled and interdependent, regardless of their physical distance. This effect is frequently referred to as "spooky action at a distance," because changes to one entangled qubit instantly alter the state of the other, regardless of spatial separation. Entanglement is a valuable resource for quantum computation, allowing the transmission of information and correlations that classical systems cannot.

Quantum Gates and Quantum Circuits:

Quantum gates serve as the foundation for quantum circuits, which manipulate qubits to execute calculations. Quantum gates are similar to conventional logic gates like AND, OR, and NOT gates in that they work on qubits in a quantum context. Quantum gates may execute operations like as rotations, flips, and entanglements, allowing qubit states to be transformed and manipulated. Quantum circuits can perform sophisticated computations and algorithms by integrating numerous quantum gates in a specified order.

Quantum algorithms are built with quantum circuits made up of numerous quantum gates. These algorithms use qubits' unique qualities, such as superposition and entanglement, to solve computing problems faster than traditional algorithms.

Quantum Algorithms

When compared to traditional computers, quantum computing has the potential to significantly accelerate the solution of computational problems. Now we explore some of the most prominent quantum algorithms that demonstrate quantum computing's strength and capabilities.

Shor's Algorithm:

Developed by Peter Shor in 1994, Shor's Algorithm is a revolutionary quantum algorithm for factoring huge integers. Large numbers must be factored into their prime factors, which is a computationally difficult job that is at the heart of many cryptographic systems. Shor's algorithm uses quantum features like superposition and entanglement to factor big numbers substantially quicker than traditional algorithms. Shor's algorithm has significant implications for cryptography, as it poses a substantial threat to widely used encryption methods like RSA, which rely on the difficulty of factoring big numbers.

Grover's Algorithm:

Grover's algorithm is a quantum search algorithm invented by Lov Grover in 1996 that delivers a quadratic speedup over traditional search methods. In comparison to the linear time required by traditional search algorithms, it can efficiently explore an unsorted database of N items in around $\sqrt{N}$ iterations. Grover's algorithm accomplishes this speedup through the use of quantum phenomena such as superposition and interference. Grover's algorithm has major implications for applications such as database searching, optimisation issues, and constraint fulfilment, despite the fact that the speedup is not exponential like Shor's algorithm.

Other Quantum Algorithms:

In addition to Shor's and Grover's algorithms, several other quantum algorithms have received interest due to their possible applications. The Quantum Fourier Transform (QFT) is a fundamental quantum computer method that is used in many quantum algorithms, including Shor's algorithm. QFT allows for the effective manipulation of quantum states in the frequency domain and serves as the foundation for many quantum signal processing applications.

Another famous quantum protocol that handles difficulties in quantum chemistry and material science is the Variational Quantum Eigensolver (VQE). VQE uses quantum computing capabilities to estimate the ground-state energy of molecules, providing insights into chemical reactions and material characteristics. VQE seeks the lowest energy state of a molecular system by optimising the parameters of a variational ansatz.

Quantum Hardware

The field of quantum computing comprises a wide range of hardware platforms that serve as the foundation for implementing quantum computation. Now we see some of the most important forms of quantum hardware and the difficulties connected with their development.

Superconducting Qubits:

Superconducting qubits have been the focus of major research and development as a prominent platform for quantum computing. Superconducting circuits are commonly used to implement them, with the qubits represented by the quantum states of electrical circuits. Scalability and the capacity to perform high-fidelity operations are two advantages of superconducting qubits. However, they confront difficulties with qubit coherence, which refers to a qubit's capacity to preserve its quantum state in the presence of ambient noise.

Trapped Ions:

Trapped ion systems are another popular quantum computing platform. Using electromagnetic fields, qubits are encoded in the internal energy levels of trapped ions. Because trapped ions have lengthy coherence durations and high gate fidelities, they are suitable for conducting complex quantum processes. However, because to the precise control necessary for controlling individual ions, scaling trapped ion systems to a high number of qubits remains a challenge.

Topological Qubits:

Topological qubits are a theoretical approach to quantum computing that involves encoding and manipulating qubits using topological states of matter. These qubits are naturally resistant to certain types of errors, allowing for fault-tolerant quantum processing. The practical realisation of topological qubits, on the other hand, is still an active topic of research, and considerable technological advances are necessary to bring them to completion.

Photonic Qubits:

Photonic qubits utilise photons as quantum information carriers. Optical components such as beamsplitters and wave plates can be used to alter them. Photonic qubits have benefits such as reduced decoherence rates and the ability to transport data across great distances via optical fibres. However, implementation of efficient photon sources and detectors, as well as executing high-fidelity operations on photonic qubits, remain problems.

There are various hurdles to building practical and scalable quantum computers. One of the most difficult issues is achieving lengthy qubit coherence periods, as ambient noise and interactions with the surrounding environment can cause quantum states to degrade. Furthermore, increasing the number of qubits while maintaining their entanglement presents substantial technical challenges.

Researchers and technologists are working hard to address these issues. Materials research, fabrication processes, and error correction codes are all improving qubit coherence times and fidelity. Efforts are also being made to create hybrid techniques that combine multiple types of qubits to maximise their respective capabilities.

Quantum Error-Correction

Quantum error correction is a key concept in quantum computing that tries to reduce the negative effects of noise and decoherence in quantum systems. Now we explore the basics of quantum error correction and the approaches used to safeguard quantum information.

The Need for Quantum Error Correction:

Qubits in quantum systems are subject to a variety of noise sources, including environmental interactions, control mistakes, and flaws in hardware components. These noise sources can cause mistakes to propagate via the quantum circuit, resulting in coherence loss and quantum information corruption. Quantum error correction detects and corrects these errors, preserving quantum state integrity and enabling reliable quantum computation.

Quantum error correction codes are specialised encoding systems that continuously spread quantum information across multiple qubits. By encoding logical qubits into a higher number of physical qubits, these codes introduce redundancy. The redundancy enables the detection and correction of errors throughout the quantum computation process.

Stabiliser Codes:

Stabiliser codes are a popular type of quantum error correcting code. Stabiliser codes employ a collection of stabiliser operators that communicate with one another and can be measured to detect errors. It is possible to determine the incidence of errors and infer the corrections required to recover the stored quantum information by measuring these stabiliser operators. The well-known three-qubit and five-qubit codes are examples of stabiliser codes.

Quantum error correction techniques involve encoding the logical qubits into physical qubits using specialised quantum error correction codes. The encoding procedure introduces redundancy, allowing errors to be detected by condition measurements. The stabiliser operators associated with the code are measured to offer information about the error conditions encountered in the system. Error correction operations can be used to restore the encoded information based on these conditions.

The implementation of quantum error correction is not without difficulties. Error correction codes enhance redundancy at the expense of increased resource requirements, such as the need for extra physical qubits and additional computational overhead. As a result, implementing fault-tolerant quantum error correction is a difficult and time-consuming operation.

Scalability is a critical factor in quantum error correction. The number of physical qubits required for error correction grows exponentially as the number of qubits in a quantum computer grows. Furthermore, the system's intrinsic gate and measurement mistakes might build during error correction, thereby undermining the effectiveness of error correction codes.

Efforts are being made to create more efficient and fault-tolerant error correcting codes that strike a compromise between the amount of redundancy necessary and the resources required for implementation. The goal of quantum error correction research is to develop codes with improved error detection capabilities, greater fault-tolerant thresholds, and lower overhead.

Applications

Quantum computing has enormous potential to transform several fields by solving problems that are currently intractable for traditional computers. Now, we will look at some of the most important applications of quantum computing and the influence it can have on several sectors.

Cryptography:

One of the most visible uses of quantum computing is in the field of cryptography. By effectively factoring big numbers, quantum algorithms such as Shor's algorithm have the potential to break widely used public-key encryption systems such as RSA and ECC. This offers a huge challenge to current cryptographic systems, triggering the development of post-quantum cryptography, which tries to create quantum-resistant encryption algorithms. Quantum computing also opens up new avenues for secure communication, such as quantum key distribution, which uses quantum entanglement principles to create unbreakable encryption keys.

Quantum Simulation:

Another possible application of quantum computing is quantum simulation. Quantum computers can represent and simulate complicated quantum systems that are beyond the capabilities of classical computers by using quantum mechanics principles. This has significant implications in domains as diverse as chemistry, materials science, and drug discovery. Quantum simulations can provide information on the behaviour of molecules, chemical reactions, and material properties, allowing for the development of new pharmaceuticals, catalysts, and materials with improved properties.

Optimisation and Machine Learning:

Quantum computing has the potential to transform optimisation problems by giving exponential speedups over traditional approaches. Many real-world problems, such as supply chain optimisation, portfolio optimisation, and scheduling, require selecting the best solution from a large set of options. Quantum techniques, such as the Quantum Approximate Optimisation Algorithm (QAOA) and the Quantum Annealing Algorithm, can improve efficiency and allow for more effective solutions to certain optimisation difficulties. Furthermore, quantum machine learning techniques are being investigated, with the goal of enhancing the capabilities of classical machine learning algorithms by using quantum algorithms and quantum data representations.

Other Applications:

Aside from cryptography, simulation, and optimisation, quantum computing has shown promise in a variety of other fields. In physics and fundamental research, for example, quantum computers can help solve complex quantum mechanical issues, explore quantum field theories, and investigate the behaviour of quantum systems. Quantum computers could also help with financial modelling, data analysis, weather forecasting, and logistics optimisation. Furthermore, quantum computing has the potential to advance artificial intelligence development by providing more efficient algorithms for applications such as pattern recognition and natural language processing.

Future

Although quantum computing is still in its early phases, it holds huge potential for the future. We are able to predict numerous significant developments and difficulties as researchers and engineers continue to push the boundaries of quantum technology.

Improving Qubit Coherence:

Maintaining the coherence of qubits—the delicate quantum states that form the basis of computation—is one of the most difficult tasks in quantum computing. Qubits are subject to noise and decoherence, which can cause errors and limit quantum system scalability. Future advances in qubit designs, materials, and production techniques will strive to improve qubit coherence, enabling longer computation times and more complicated computations. Furthermore, creating error mitigation measures and error-resistant quantum algorithms will be critical in overcoming the qubit coherence restrictions.

Advancements in Error Correction:

In quantum systems, quantum error correction is critical for reducing the effects of noise and mistakes. The number of qubits and the complexity of operations increase as quantum computers scale up, making error correction approaches vital. Future research will concentrate on enhancing and inventing new error correction codes that can handle higher levels of noise and errors while using fewer resources. Fault-tolerant quantum computing advances will be critical for establishing error-corrected, reliable quantum computation.

Scalling up quantum systems:

Another significant challenge for the future of quantum computing is scaling up quantum systems. Presently, quantum computers with a few dozen qubits are available, but to realize the full potential of quantum computing, we need to scale up to hundreds, thousands, or even millions of qubits. This requires significant advancements in quantum hardware, such as improving qubit connectivity, reducing crosstalk, and increasing the number of qubits while maintaining their coherence. Developing scalable architectures and engineering solutions will be critical for building large-scale, fault-tolerant quantum computers.

3.2 Quantum Cryptography

Introduction

This section explores interesting area of quantum cryptography, which uses quantum mechanics principles to establish secure communication.

We investigate the flaws of traditional encryption systems and how quantum cryptography addresses these issues. We investigate several facets of quantum cryptography, from fundamental principles to practical implementations.

The requirement for safe information transmission has grown increasingly important in modern communication. For decades, traditional cryptography, which depends on mathematical methods and keys, has been the bedrock of secure communication. However, as computational power increases, the flaws of traditional encryption systems become increasingly evident.

For their security, traditional cryptographic systems, such as the widely used RSA and AES algorithms, rely on the complexity of particular mathematical problems. The security of RSA encryption, for example, is dependent on the computational difficulties of factoring huge integers. The development of powerful quantum computers, on the other hand, poses a significant risk to existing traditional cryptography methods.

Quantum computers, which use quantum physics concepts, have the ability to answer certain mathematical problems far quicker than classical computers. Shor's algorithm, a quantum algorithm specifically intended to factor huge numbers, reveals quantum computers' capacity to efficiently breach the widely used RSA encryption scheme. This realisation raises questions about classical cryptography systems' long-term security in the face of quantum threats.

Researchers have looked to quantum cryptography as a possible approach to overcome these issues. Quantum cryptography secures communication by utilising fundamental laws of quantum mechanics. It uses principles like the uncertainty principle, superposition, and entanglement to create secure communication channels that are impervious to interception and tampering.

The quantum cryptography concept represents a paradoxical change in secure communication. Unlike traditional cryptographic approaches that rely on mathematical complexity, quantum cryptography uses physical rules to ensure the security of information transmission. Quantum cryptography, by utilising quantum mechanics principles, has the ability to provide provably secure communication channels that are impervious to attacks from both classical and quantum computers.

Quantum Key Distribution

Quantum key distribution (QKD) is a new cryptographic scheme that uses quantum mechanics principles to generate unbreakable encryption keys. Unlike traditional cryptography approaches, which rely on computing complexity, QKD relies on fundamental principles of quantum mechanics to assure key exchange security.

The first quantum key distribution protocol, known as BB84, was proposed in 1984 by physicists Bennett and Brassard. This technique established the groundwork for safe key exchange using quantum concepts. Since then, numerous new protocols, such as Ekert's E91 protocol, have been devised, each with its own set of strengths and weaknesses.

The Bennett-Brassard 1984 (BB84) protocol involves the exchange of quantum states, known as qubits, between a sender (Alice) and a receiver (Bob). The qubits are constructed in one of two orthogonal bases, which are commonly represented as 0 and 1. Alice encodes each bit of the key she wishes to share with Bob in one of the two bases at random. Bob, who is uninformed of Alice's basis selection, measures the received qubits at random in one of the two bases as well. Following the transmission, Alice and Bob interact publicly in order to compare a subset of their measurement findings, excluding bits measured in different bases. The remaining bits combine to generate a safe key that only Alice and Bob know.

The security of QKD protocols is based on quantum physics principles. The Heisenberg uncertainty principle, in particular, ensures that any attempt to eavesdrop or measure the qubits during transmission would disrupt their quantum states. Alice and Bob would perceive this disturbance, alerting them to the existence of an eavesdropper. As a result, QKD provides a method for detecting and preventing eavesdropping attempts, ensuring the confidentiality of the encryption keys.

The E91 protocol, invented by Ekert in 1991, uses the entanglement phenomena to provide safe key distribution. Alice and Bob share pairs of entangled particles known as Bell pairs that are in an entangled superposition of states in this protocol. Alice and Bob can create a secure key based on the correlations discovered in their measurement findings by measuring their respective particles. Because the particles are entangled, any effort to intercept the signal will disrupt the correlation, alerting Alice and Bob to the presence of an eavesdropper.

QKD protocol development and deployment have resulted in substantial advances in secure communication. QKD provides a secure means for exchanging encryption keys, as any attempt to intercept or tamper with the communication may be detected. This trait makes QKD particularly appealing for high-security applications such as government communications, financial transactions, and sensitive data transfer.

It is crucial to highlight, however, that QKD algorithms are not without challenges. Practical implementations suffer by challenges such as qubit loss, noise, and channel constraints, all of which can reduce the efficiency and range of secure key distribution. Attempts are being made to solve these issues and increase the scalability and reliability of QKD systems.

QKD Implications

Quantum Key Distribution (QKD) can be implemented using a variety of quantum devices as cryptographic key carriers. Each implementation has its own set of benefits and drawbacks, and academics have made tremendous progress in establishing viable QKD systems. Now, we will look at some of the most important implementations and their properties.

One common implementation of QKD involves the use of single photons as cryptographic key carriers. The qubits are typically encoded in the polarisation states of individual photons in this manner. The sender (Alice) prepares photons in certain polarisation states, indicating the BB84 protocol's encoding of 0 and 1. Using appropriate detectors, the receiver (Bob) determines the polarisation of the received photons. Alice and Bob can create a secure key by comparing their measurement findings for a subset of the photons.

Using single photons to implement QKD poses various obstacles. The detection efficiency of photon detectors is a key factor. To ensure accurate measurement and to reduce the impacts of noise, high-efficiency detectors are required. Furthermore, the transmission line between Alice and Bob can incorporate a variety of noise sources, such as background light and losses. To address these issues, techniques such as active compensation and error correction methods are used.

Another QKD method makes use of entangled photon pairs, such as those produced via spontaneous parametric down-conversion. In this method, Alice and Bob share a pair of entangled photons. Entanglement enables the creation of secure keys based on correlations discovered in their measurements. Alice and Bob can extract a secure key from the entangled photon pairs by modifying the measurement bases and executing appropriate measurements.

Entanglement-based QKD implementations present unique challenges. It is a difficult to generate and distribute entangled photon pairs over vast distances while maintaining the quality of entanglement. Noise and decoherence can further degrade the entangled state's quality, necessitating the adoption of error correction techniques to assure secure key creation.

Other quantum systems have been investigated for QKD implementations, in addition to single photons and entangled photon pairs. Other types of qubits, such as superconducting circuits, trapped ions, or quantum dots, are used in these applications. Each system has its own set of characteristics and challenges, and continuing research is aimed at improving their performance for secure key distribution.

To show the practicality and possible applications of QKD systems, they have been constructed and incorporated into existing communication networks. Additional components, such as traditional communication routes, synchronisation methods, and key distillation techniques, are frequently used in these systems. Integration with current infrastructure enables QKD to be seamlessly included into traditional communication protocols, offering a higher level of security for sensitive data transmission.

It should be noted that real QKD implementation still confronts hurdles in terms of scalability, cost, and compatibility with existing network infrastructures. Overcoming these obstacles will necessitate technological advances, such as the creation of more efficient and reliable quantum systems, developments in detection techniques, and the incorporation of QKD into larger-scale networks.

Quantum Cryptographic Protocols

In addition to Quantum Key Distribution (QKD), there are various quantum cryptography methods that use quantum mechanics concepts to secure communication. These protocols provide unique security features that traditional cryptography approaches cannot supply. Two significant examples are discussed here: Quantum Secure Direct Communication (QSDC) protocols and Quantum Coin Flipping.

Quantum Secure Direct Communication (QSDC) protocols allow for the direct and secure exchange of data between two parties without the use of shared encryption keys. Unlike typical communication techniques that send encrypted messages over a public channel, QSDC protocols use quantum principles to maintain the confidentiality and integrity of the sent data.

The sender (Alice) encodes the information she wishes to broadcast into a series of quantum states in a standard QSDC protocol. These states are subsequently sent through a quantum channel to the receiver (Bob). Bob extracts the encoded information by performing appropriate measurements on the received states. Importantly, any eavesdropping efforts or unauthorised measurements made by an adversary (Eve) would cause the quantum states to be disrupted, revealing the presence of an interception.

QSDC procedures have a number of advantages. They enable immediate and secure communication without the need to first establish and exchange encryption keys. Furthermore, QSDC protocols provide information-theoretic security, which means that the security of the communication is assured by quantum mechanics laws.

Quantum Coin Flipping is another fascinating quantum cryptography scheme. Coin flipping is a fundamental operation in cryptography in which two parties, known as the sender and receiver, desire to generate a random outcome without trusting each other. To provide a fair and secure coin flipping process, quantum coin flipping methods employ quantum principles.

The transmitter and receiver both contribute quantum states and make measurements on them in a Quantum Coin Flipping mechanism. The ultimate result is determined by the measurement data as well as a specified rule. The security of Quantum Coin Flipping protocols is based on the notion that any attempt by one party to influence the outcome or cheat will be noticed by the other side due to quantum mechanics laws.

In cases where trust between parties is restricted or nonexistent, quantum coin flipping algorithms offer intriguing applications. They can, for example, be used to establish a fair random selection of leaders in distributed systems or to provide fair gambling between participants situated in different locations.

Furthermore, Quantum Signature algorithms offer a novel approach to digital signature security by utilising quantum mechanics principles. Quantum signatures provide information-theoretic security and are immune to fraud and tampering. They can be used to validate the validity and integrity of digital messages, confirming that they are genuine and have not been tampered with.

New protocols and applications are being investigated as quantum cryptography evolves. By utilising the power of quantum mechanics, these protocols provide novel answers to cryptography issues. These protocols provide security features that are fundamentally different from classical cryptography systems by exploiting the principles of superposition, entanglement, and quantum measurements.

Quantum Cryptanalysis

With the introduction of quantum computers, a new branch of study known as quantum cryptanalysis has emerged. The goal of quantum cryptanalysis is to create algorithms and approaches that use the capability of quantum computers to break classical cryptography systems. As discussed quantum algorithms, such as Shor's algorithm, have showed the ability to efficiently factor big numbers and solve key mathematical problems that constitute the foundation of traditional encryption systems. As a result, the security of several frequently used cryptographic techniques, such as RSA and elliptic curve cryptography, is put at risk by the advent of practical quantum computers.

The influence of quantum computers on traditional encryption systems is a major source of concern among the security community. While it would take an infeasible amount of time for classical computers to defeat current encryption techniques, quantum computers could potentially render them vulnerable. This has increased the demand for post-quantum cryptography, or the creation of cryptographic methods that remain secure even in the face of powerful quantum computers.

Post-quantum cryptography seeks to develop encryption algorithms, digital signatures, and other cryptographic primitives that are resistant to both classical and quantum computer attacks. These methods are intended to survive anticipated future advances in quantum computing while also ensuring the long-term security of sensitive data.

Several post-quantum cryptographic algorithm families have been proposed and are now being researched. These algorithms are based on mathematical issues that are thought to be difficult to solve by both classical and quantum computers. Lattice-based cryptography, code-based cryptography, multivariate polynomial cryptography, and hash-based cryptography are other examples. These algorithms rely on mathematical structures and computing issues that cannot be solved efficiently by known quantum algorithms.

The shift to post-quantum cryptography comes with difficulties. It necessitates not only the creation and regulation of new cryptographic algorithms, but also their incorporation into current systems and protocols. Post-quantum cryptography usage will necessitate a careful mix of security, performance, and compatibility with legacy systems.

Post-quantum cryptography algorithms are being evaluated and selected by standardisation agencies and research organisations. The goal is to properly analyse these algorithms and fully comprehend their security features. This procedure involves rigorous examination and competition among several proposals in order to discover the most promising and secure post-quantum cryptography algorithms.

The implementation of post-quantum cryptography will most likely include a hybrid technique that combines conventional and post-quantum algorithms. This enables for a gradual transition while maintaining compatibility with existing systems. Cryptographic protocols and systems will need to be upgraded to support the incorporation of post-quantum algorithms, and cryptographic key management practises will need to be improved to fit the new requirements.

As the science of quantum cryptanalysis and post-quantum cryptography evolves, it is critical for researchers, industry experts, and regulators to interact and keep updated. The development and deployment of post-quantum cryptography solutions will be crucial in ensuring the security and privacy of sensitive information in a future where powerful quantum computers become a reality.

Challenges and Future Directions

While quantum cryptography holds immense potential for secure communication, a number of difficulties and research topics must be solved before its full potential can be realised. Now, we will look at some of the important difficulties and future directions in quantum cryptography.

Limitations with QKD

Current quantum key distribution (QKD) systems have practical limitations that must be overcome before they can be widely adopted. One such limitation is the distance that safe key distribution can be accomplished over. The maximum distance between communicating parties is limited by photon loss during transmission. To improve the range of QKD systems, researchers are actively investigating approaches such as quantum repeaters and quantum memory. Furthermore, noise, such as background noise and detector inefficiencies, offers a substantial issue, requiring the development of effective error correction and privacy amplification approaches.

Scalability and efficiency are two additional challenges in quantum cryptography. As the number of users and the complexity of cryptographic protocols grow, developing scalable quantum cryptography systems becomes critical. This includes effective key distribution procedures, quick and dependable key production, and safe key management. Improving the effectiveness of quantum cryptography protocols will allow them to be integrated into current communication networks and ensure their use in real-world applications.

With the possibility of quantum computers breaching traditional encryption systems, the creation of post-quantum cryptographic algorithms has become a critical research direction. The goal of quantum-resistant cryptography is to offer strong encryption algorithms that are resistant to attacks from both conventional and quantum computers. Ongoing research is centred on the study and creation of post-quantum cryptography algorithms that provide strong security guarantees while remaining economical and practical to deploy.

Integration with other emerging quantum technologies can benefit quantum cryptography. Quantum networks, which allow quantum information to be transmitted between nodes, can extend the scope and capability of quantum cryptography algorithms. The quantum internet, a worldwide network of interconnected quantum nodes, has the potential to enable secure global communication. The combination of quantum cryptography, quantum networks, and the quantum internet opens up new opportunities for safe and efficient communication in the quantum age.

As quantum cryptography systems advance in commercialization, the development of standards and certifications becomes critical. To ensure interoperability, security, and dependability across diverse quantum cryptography methods and systems, standardisation initiatives are required. Certification processes will give users and organisations with assurance about the security and performance of quantum cryptography technologies.

Quantum hacking and countermeasures:

The field of quantum hacking is evolving in combination with quantum cryptography. Researchers are currently investigating potential flaws and creating countermeasures to attacks on quantum cryptography systems. This includes researching side-channel threats, quantum hacking techniques, and creating safe hardware and software implementations.

3.3 Quantum Communication

Introduction

This section explores the fascinating field of quantum communication, which uses quantum physics principles to ensure safe and efficient information transmission.

Quantum Networks

Quantum networks are an essential component of quantum communication infrastructure, allowing quantum information to be transmitted between different nodes. These networks allow for the propagation of entangled states, quantum state teleportation, and the execution of quantum protocols over long distances. Let's look at quantum networks' architecture, components, problems, and applications.

Components and Architecture: Quantum networks are made up of interconnected nodes that communicate quantum information. Each network node can serve as a source, receiver, or relay of quantum states. Quantum network architecture can vary based on the application and physical implementation.

Quantum Repeaters: Quantum repeaters are an important component of quantum networks that are meant to overcome the constraints of quantum signal loss across long distances. To extend the range of entanglement while maintaining its fidelity, they use entanglement swapping and entanglement purification techniques. Quantum repeaters enable reliable transmission of quantum information over greater distances by splitting the network into shorter segments.

Quantum Memories: Quantum memory are technologies that can store and retrieve quantum states. They are critical in quantum networks because they enable synchronisation and synchronization-free communication between nodes. Quantum memories allow quantum information to be stored until it is needed for processing or transfer.

Quantum Routers: Quantum routers direct quantum states to their intended destinations by managing the flow of quantum information inside a network. They can route quantum signals based on their destination, prioritise specific forms of quantum information, and control network resource allocation.

The challenges of Developing Quantum Networks:

There are various problems to building and scaling quantum networks that must be addressed:

Maintaining Entanglement: Maintaining entanglement over long distances is one of the most perplexing challenges. Quantum signals can be degraded by many types of noise and environmental interactions. To overcome these obstacles and increase the extent of entanglement, quantum repeaters and error correction techniques are used.

Synchronising Quantum processes: Coordinating quantum processes across numerous nodes is critical for quantum network success. To ensure proper information flow and processing, accurate synchronisation of operations such as measurements and entanglement production is required.

Managing Network Resources: Effectively managing network resources such as entangled states, memory capacity, and bandwidth is critical for quantum network scalability. To optimise network resource utilisation, strategies for resource allocation, network routing, and dynamic adaptability are being developed.

Quantum Communication for Quantum Computing

Quantum communication is critical in the science of quantum computing because it allows quantum states to be transmitted across different components of a quantum computer. Let's look at the importance of quantum communication in quantum computing and some essential features of this topic.

Distributed Quantum Computing: Quantum computing frequently involves distributed systems with physically separated quantum processors, quantum memory units, and other components. Quantum communication is required in such settings to transport quantum states between different components of the quantum computer. For example, quantum gates may work on qubits in distinct physical locations, necessitating their entanglement via quantum communication. Because quantum computing is distributed, it allows for parallel processing, collaboration among different quantum computers, and the possibility of tackling complicated problems more effectively.

Quantum Communication Requirements in Quantum Computing:

In the context of quantum computing, quantum communication imposes particular requirements on communication protocols and channels. Among the major considerations are:

Low Error Rates: Quantum states are vulnerable to noise and decoherence. To retain the authenticity of transmitted quantum states, quantum communication protocols for quantum computing systems should aim towards low error rates.

High Fidelity: Maintaining high fidelity in quantum communication is critical to ensuring reliable quantum information transfer. Any loss or distortion during communication might decrease the quality of the quantum states and harm computing.

Low Latency: Quantum communication protocols should have low latency to enable efficient and real-time communication between quantum computer components. Low latency is especially significant in distributed quantum computing, where timely quantum state transmission is critical for synchronisation and parallel computation.

Current Research and Challenges:

There are various hurdles to developing efficient and trustworthy quantum communication protocols for quantum computing applications. Among these difficulties are:

Quantum Channel Noise: Because of the interaction with the environment, quantum communication is prone to noise and decoherence. To overcome this difficulty, error-correcting codes, quantum repeaters, and other strategies for reducing noise and preserving quantum states during transmission must be developed.

Scalability: As quantum computing systems scale up, so do the needs for quantum communication. An active topic of research is ensuring efficient and scalable quantum communication protocols capable of handling large-scale quantum systems.

Quantum Teleportation Protocols: Quantum teleportation protocols are important in quantum communication and quantum computation. An current research focus is on developing efficient and high-fidelity teleportation systems capable of transferring quantum states over large distances.

Researchers are currently researching these issues and developing new protocols and approaches to increase the efficiency, reliability, and scalability of quantum communication for quantum computing applications. By overcoming these issues, quantum communication can enable the seamless transfer of quantum states within quantum computing systems, allowing powerful quantum algorithms and applications to be realised.

3.4 Quantum Sensors and Metrology

Introduction

Now in this section we begin by discussing the topic of quantum sensors and metrology, which is concerned with utilising quantum features for very exact and sensitive measurements. We address the constraints of traditional measuring techniques as well as the possible benefits of quantum sensors. We investigate how quantum principles such as superposition and entanglement can be used to improve measurement accuracy and precision.

Quantum Metrology

Quantum metrology is a subfield of metrology that uses quantum mechanics concepts to achieve highly precise and sensitive measurements. Quantum metrology has the ability to overcome the limits of traditional measuring techniques by utilising quantum resources such as superposition and entanglement. Now, we will look at the underlying ideas and techniques of quantum metrology, as well as some of its applications in many domains.

Heisenberg's uncertainty principle, which asserts that there is a fundamental limit to the precision with which certain pairs of physical parameters, such as location and momentum, can be measured simultaneously, is at the heart of quantum metrology. However, quantum metrology seeks to capitalise on this uncertainty in order to improve measurement precision. The quantum Cramér-Rao bound is a mathematical restriction on the precision of any unbiased estimation of a parameter that is frequently used to evaluate the effectiveness of quantum metrology protocols.

Quantum parameter estimation, which includes estimating an unknown parameter with high precision using quantum systems, is an important technique in quantum metrology. Quantum systems can be created in superposition states, allowing several parameter values to be measured at the same time. Quantum parameter estimation can reach precision beyond what is attainable with classical procedures by optimising the measurement scheme and utilising quantum algorithms.

Another effective technique in quantum metrology is quantum interferometry. It makes use of the wave-like characteristics of quantum particles like photons or atoms to measure phase changes with high precision. Interference between separate quantum particle routes results in a pattern that is extremely sensitive to even minor changes in phase. Precision measurements of physical quantities such as length, frequency, and gravitational waves are now possible. Quantum interferometry has the potential to transform precision measurements in fields like gravitational wave detection and precision timekeeping.

Squeezed states, which are a sort of non-classical quantum state, are also important in quantum metrology. These states exhibit lower fluctuations in one of the measured values, such as amplitude or phase, but higher fluctuations in the conjugate property. Squeezed states can improve measurement precision beyond what is possible with classical states by squeezing the noise in the measurement process. They have applications in domains such as magnetic field detection, which requires precise measurements of small magnetic fields.

The benefits of quantum metrology can be found in a variety of domains. Quantum metrology techniques can increase the accuracy of atomic clocks, which are critical for applications such as global positioning systems (GPS) and communication network synchronisation. Quantum metrology allows for the exact measurement of microscopic distortions in space-time generated by gravitational waves, offering up new possibilities for exploring the universe. In magnetic field sensing, quantum metrology has the ability to detect magnetic fields with great sensitivity, with applications spanning from medical imaging to materials characterisation.

Quantum Sensors

Quantum sensors are an advanced tool that uses quantum mechanics principles to monitor physical quantities with extraordinary sensitivity and precision. To overcome the constraints of conventional sensors, these sensors use quantum phenomena such as superposition, entanglement, and quantum coherence. Now we will look at many types of quantum sensors and their applications in a variety of fields.

The atom interferometer is one sort of quantum sensor that uses the wave-like features of atoms to measure physical quantities such as position, velocity, and acceleration. Atom interferometers function by employing laser beams to divide a cloud of ultra-cold atoms into two independent routes, allowing the atoms to follow both tracks at the same time. When the routes recombine, the interference pattern that occurs provides information about the quantity being measured. Atom interferometers are employed in applications such as inertial navigation systems, gravitational field mapping, and geodesy because of their high sensitivity.

Another significant variety of quantum sensors is quantum magnetometers, which measure magnetic fields with high precision. These sensors detect even extremely weak magnetic fields by utilising the quantum characteristics of atoms or solid-state systems. Atomic magnetometers, for example, detect magnetic fields by using atomic ensembles or single atoms in a superposition state. They are used in a variety of domains, including geophysics, medical imaging, and fundamental physics research.

Quantum gravimeters are sensors that measure gravitational acceleration with extreme precision. To detect minute changes in gravitational forces, they use techniques such as atom interferometry or superconducting quantum circuits. Quantum gravimeters have the potential to significantly advance geodesy, geophysics, and the investigation of fundamental physical processes like gravitational wave detection.

Quantum thermometry is a field that uses quantum systems to measure temperature with high precision. Quantum sensors can take advantage of the thermal features of quantum states or the temperature sensitivity of quantum coherence. In sectors such as materials science, cryogenics, and biological research, these sensors have the ability to give exact temperature measurements.

The benefits of quantum sensors are numerous. They have a great sensitivity, allowing them to detect extremely faint impulses that would be difficult to detect with other sensors. Quantum sensors have low noise characteristics, which reduces the impact of ambient disturbances and improves measurement accuracy. Furthermore, because quantum sensors can work at such small sizes, they are well suited for applications in nanotechnology and biology.

The applications of quantum sensors are numerous and growing. Quantum sensors are used in navigation systems, mineral exploration, environmental monitoring, and quantum information processing, in addition to the domains described above. We should expect the creation of new and innovative quantum sensors with improved performance and a larger range of applications as quantum technologies advance and become more accessible.

Applications of Quantum Sensing

Quantum sensors offer several applications in a variety of sectors, revolutionising how we measure and comprehend the physical world. Now we will look at some of the various uses of quantum sensors and their impact on various industries and scientific disciplines.

Precision navigation and inertial sensing are two important applications of quantum sensors. Quantum sensors, such as atom interferometers and quantum gyroscopes, outperform traditional sensors in terms of accuracy and stability. They find use in self-driving cars, drones, and spaceships, where precise navigation and inertial sensing are critical. Quantum sensors can improve navigation system performance by allowing for more accurate positioning, velocity readings, and attitude control.

Another area where quantum sensors have made important advances is geophysical research. Quantum sensors can provide vital insights into subsurface structure, mineral deposits, and geological events by sensing small changes in gravity, magnetic fields, or even seismic activity. These sensors are used to map underground resources, detect natural disasters and monitor environmental conditions in mineral exploration, oil and gas exploration, and environmental monitoring.

Quantum sensors have the potential to revolutionise diagnostic approaches in the world of biomedical imaging. Quantum magnetometers, for example, can precisely detect the magnetic fields created by the human body, enabling extraordinarily sensitive imaging of the brain and other organs. Quantum sensors can also be used to monitor physiological processes such as heart and muscle activity in a non-invasive manner, providing significant information for medical diagnosis and study.

Quantum sensors are essential for the advancement of quantum technologies such as quantum communication, quantum computation, and quantum internet. Precise measurements are required in quantum communication for reliable transfer of quantum states and secure key distribution. Quantum sensors collect and characterise quantum signals, which ensures the integrity and security of quantum communication protocols.

Quantum sensors are used in quantum computing to measure and manipulate individual qubits, which are the fundamental units of quantum information. Controlling and measuring qubits precisely is critical for executing quantum processes and receiving correct results. Quantum sensors are essential for characterising and calibrating quantum systems, which helps to construct scalable and error-corrected quantum computers.

Furthermore, quantum sensors are used to assess and monitor the quality of quantum connections in the growing field of quantum internet. These sensors identify and fix errors that occur during the transmission of quantum data across great distances. Quantum sensors are essential for creating trustworthy quantum links and providing secure and efficient quantum communication inside a quantum network.

Beyond the examples given above, the applications of quantum sensors are numerous. They are used in metrology, materials science, environmental monitoring, and fundamental physics research. We may expect quantum sensors to play an increasingly important role in precise measurements, scientific discoveries, and technological advances as quantum technologies progress and become more accessible.

Chapter 4

The Quantum World of Particles

4.1 Standard Model of Particle Physics

Introduction

This section explores the Standard Model of particle physics, the dominant hypothesis that defines fundamental particles and their interactions. We investigate the Standard Model's historical evolution and important components. We reveal the complicated web of particles and forces that regulate the tiny world, from quarks and leptons to gauge bosons and the Higgs boson.

Elementary Particles

The basic particles, which are the fundamental building elements of matter, are at the centre of the Standard Model of particle physics. These particles are divided into two groups: quarks and leptons.

Quarks:

Quarks are fundamental particles with fractional electric charges that make up protons and neutrons, the particles that make up the atomic nucleus. Quarks are classified into six sorts, each having its unique mass and charge: up (u), down (d), charm (c), strange (s), top (t), and bottom (b). The lightest and most frequent quarks in ordinary matter are the up and down quarks, which have charges of $+2/3e$ and $-1/3e$, respectively. The remaining four quarks are more heavier and less stable, and can only be found in high-energy situations like particle colliders or the early cosmos.

Leptons:

Leptons are another type of elementary particle that does not experience the strong nuclear interaction and is not made up of quarks. The most well-known lepton is the electron (e), which has a negative electric charge and orbits the atomic nucleus. The electron has a substantially lower mass than quarks, making it significantly lighter. There are two more charged leptons: the muon (μ) and the tau (τ), which are substantially heavier and have shorter lives. Along with these charged leptons, the family of leptons is completed by their equivalent neutrino (ve, vμ, and vτ), which is electrically neutral and nearly massless.

Properties and Interactions:

The properties of these fundamental particles define their behaviour and interactions. Particles have three fundamental properties: mass, charge, and spin. A particle's mass determines the quantity of matter it contains, its charge refers to the electric charge it carries, and its spin quantifies its intrinsic angular momentum. The interactions that regulate particle behaviour are mediated by the fundamental forces: electromagnetic, weak nuclear, and strong nuclear forces. Quarks and leptons interact and exchange force-carrying particles known as gauge bosons.

Photons carry the electromagnetic force, which is responsible for interactions between charged particles such as electrons and protons via electromagnetic fields. Radioactive decay and neutrino interactions are caused by the weak nuclear force, which is mediated by the W and Z bosons. The strong nuclear force, mediated by gluons, is responsible for the stability of atomic nuclei by binding quarks together to form protons and neutrons.

Formation of Matter:

The interaction of quarks and leptons allows for the creation of atoms, which are the fundamental units of matter. Protons and neutrons, which are made up of quarks, live in the atomic nucleus and are held together by the strong nuclear force. Electrons, on the other hand, are held in place by the electromagnetic force in the electron shells around the nucleus. The chemical properties and behaviour of distinct elements in the periodic table are determined by the arrangement of these particles.

Diversity of matter:

The vast diversity of matter observable in the universe is due to the many combinations of quarks and leptons. Different forms of hadrons, including protons and neutrons, are produced by organising quarks in various ways. After that, hadrons mix with electrons to form atoms. From the smallest subatomic particles to the enormous structures of galaxies, the unique properties and interactions of basic particles ultimately determine the features of matter.

To summarise, quarks and leptons are the fundamental constituents of matter, and their properties and interactions are critical in constructing the universe as we know it. They combine to form atoms, which give rise to the great variety of matter and the rich intricacy of the physical universe. Understanding the properties and behaviours of these elementary particles is critical to solving cosmic mysteries and expanding our understanding of fundamental physics. Experiments at high-energy particle colliders, such as the Large Hadron Collider (LHC), continue to advance our understanding of these basic particles and their interactions.

Forces and Interactions

There are several fundamental forces and interactions that regulate the behaviour of elementary particles in the Standard Model of particle physics. These forces are fundamental in shaping the universe and determining matter's properties. Let's take a closer look at them:

Electromagnetic Force: The electromagnetic force is responsible for charged particle interactions. Quantum electrodynamics (QED), a quantum field theory, describes it. Photons, which are massless particles that operate as carriers of the electromagnetic field, mediate the electromagnetic force. Charged particles, such as electrons and protons, interact with one another via photon exchange, producing attraction or repulsive forces. The electromagnetic force is important in a variety of phenomena, including electron-atomic nuclei interactions, chemical bond formation, and light propagation.

Strong Nuclear Force: The strong nuclear force, also known as the strong interaction or the strong force, is the most powerful of the four fundamental forces. It is in charge of joining quarks to produce protons and neutrons, which are known as hadrons. The strong nuclear force is carried by gluons, which operate as intermediaries between quarks. The strong nuclear force, unlike the electromagnetic force, remains constant across relatively short distances, resulting in the confinement of quarks within hadrons. The strong force is required for the stability of atomic nuclei and is important in nuclear reactions and particle decays.

Weak Nuclear Force: The weak nuclear force, commonly known as the weak interaction, is responsible for phenomena such as subatomic particle decay and quark transition. The theory of electroweak interaction, which unites the weak and electromagnetic forces, describes it. W and Z bosons are particles that mediate the weak nuclear force. It is distinguished by its short-range nature and its participation in the breaking of particle-antiparticle symmetry. The weak force causes events like beta decay, which occurs when a neutron decays into a proton, producing an electron and an electron antineutrino.

Gravity: Gravity is the force that causes objects with mass or energy to attract one another. It is described by Albert Einstein's theory of general relativity. Gravity is not considered a fundamental force in general relativity, but rather the curvature of space-time caused by the presence of mass and energy. Regardless of charge or other qualities, gravitational force acts on all particles. While the other fundamental forces have insignificant impact on macroscopic objects, gravity is responsible for celestial body behaviour, galaxies' structure, and the overall dynamics of the universe.

Unification of Forces:

One of the goals of modern physics is to combine the fundamental forces into a single, more comprehensive theory. Within the framework of quantum field theory, the Standard Model successfully describes the electromagnetic, strong nuclear, and weak nuclear forces. It does not, however, involve gravity. The search for a unified theory that incorporates all of the fundamental forces and their interactions is an ongoing scientific issue. String theory and quantum gravity, for example, seek to provide a unified description of all the forces and particles in the universe, including gravity.

In summary, the Standard Model of particle physics includes the electromagnetic force, the strong nuclear force, and the weak nuclear force, which govern the behaviour of elementary particles collectively. These forces are responsible for the tiny interactions and behaviours seen. Although gravity is not included in the Standard Model, it is a basic force that works on greater scales and is described by general relativity theory. Understanding the fundamental features of matter and the dynamics of the universe requires an understanding of these forces and their interactions. The search for a unifying theory drives scientific research and deepens our understanding of the underlying nature of reality.

Gauge Bosons and Interactions

The fundamental forces are mediated by particles known as gauge bosons in the Standard Model of particle physics. These gauge bosons are in charge of transferring energy and momentum between interacting particles, hence mediating forces and interactions. Let's look at some of the most important gauge bosons and their interactions:

Photon:

A photon is a gauge boson that is related with electromagnetic force. It is a massless particle that interacts with charged particles and carries electromagnetic force. Photons are responsible for the transmission of electromagnetic radiation, including light and other types of electromagnetic waves. Through the exchange of virtual photons, they facilitate interactions between charged particles such as electrons and protons. This interaction produces the attraction or repulsive forces seen in electromagnetic interactions.

Gluon:

The gauge boson associated with the strong nuclear force is the gluon. Gluons, unlike photons, have an electrically neutral colour charge and interact with quarks and other gluons. Gluons are important in binding quarks together to form hadrons like protons and neutrons. The strong nuclear force, mediated by gluons, is responsible for atomic nuclei's stability and the confinement of quarks within hadrons. The theory of quantum chromodynamics (QCD) describes the interactions of gluons.

W and Z Bosons:

The gauge bosons linked with the weak nuclear force are the W and Z bosons. They are heavier particles than photons and gluons and are in charge of mediating the weak interactions between elementary particles. The W bosons are charged in two forms, W+ and W-, but the Z boson is electrically neutral. These bosons are involved in processes like beta decay, which occurs when a neutron decays into a proton, releasing a W-boson, which then decays into an electron and an electron antineutrino. The low range of the weak nuclear force, as well as its violation of parity and charge-parity (CP) symmetry, characterise it.

Gauge Symmetry:

The concept of gauge symmetry is closely linked to the existence of gauge bosons and their interactions. A mathematical characteristic of a physical theory that ensures its invariance under specific transformations is referred to as gauge symmetry. In the Standard Model, gauge symmetry is connected to charge conservation, such as electric charge and colour charge. The electromagnetic, weak, and strong forces all have local gauge symmetry, which gives rise to the corresponding gauge bosons and regulates their interactions with matter particles.

Understanding the role of gauge bosons allows us to better understand the nature of fundamental forces and their interactions as described by the Standard Model. The exchange of these gauge bosons between particles allows forces to be transmitted and different physical phenomena to emerge. The study of gauge bosons and their properties is critical for examining fundamental natural principles and progressing our understanding of the cosmos at its most fundamental level.

Experimental Evidence

Various experiments and observations have extensively tested and confirmed the Standard Model of particle physics. These experimental studies give solid evidence for the model's validity and accuracy. Let us look at some of the most important pieces of experimental evidence that support the Standard Model:

High-Energy Particle Accelerators: The high-energy particle accelerator is a crucial tool in experimental particle physics. Accelerators, such as CERN's Large Hadron Collider (LHC), enable scientists to collide particles at extremely high energies in order to recreate the circumstances that prevailed right after the Big Bang. These collisions create an abundance of data that can be analysed to examine particle characteristics and evaluate Standard Model predictions.

Several particles predicted by the Standard Model have been identified by LHC experiments, notably the Higgs boson in 2012. The discovery of the Higgs boson was an important turning point in physics since it verified the process that gives mass to other particles and provided experimental proof for the Higgs field.

Neutrino Oscillations: Another major scientific observation that supports the Standard Model is neutrino oscillations. Neutrinos are elusive particles with extremely minuscule masses that interact with matter only weakly. Scientists have discovered that neutrinos can change properties as they travel through underground facilities, such as the Super-Kamiokande and IceCube experiments, showing that they have non-zero masses. This neutrino oscillation phenomena is consistent with the Standard Model's theoretical framework, which has been extended to incorporate neutrino masses.

Precision Measurements: To test the Standard Model's predictions, precision measurements of numerous particle properties were performed. Experiments at particle colliders and other high-energy facilities have yielded remarkably accurate measurements of particle masses, charges, and decay rates. These measurements have repeatedly agreed with the Standard Model's predictions, providing substantial support of its validity.

Furthermore, cosmological data such as cosmic microwave background radiation and the abundance of light elements in the universe are compatible with the Standard Model's predictions.

While the Standard Model has been remarkable in understanding fundamental particles and their interactions, it is crucial to emphasise that there are issues and events that fall outside of its purview. These include the genesis of neutrino masses, the nature of dark matter, and the unification of the fundamental forces.

Finally, experimental data from high-energy particle accelerators, precise measurements, and discoveries of phenomena such as neutrino oscillations verifies the Standard Model's predictions. The consistency of agreement between theory and experiment gives confidence in the model's accuracy and serves as a platform for further research and investigation into the underlying nature of the cosmos.

4.2 Elementary Particles and Forces

Introduction

We will look at elementary particles and the fundamental forces that control their interactions in this section. We look at the fundamental building blocks of matter, elementary particle attributes, and the four fundamental forces described by the Standard Model.

Quarks and Leptons

As mentioned earlier, Quarks and leptons are the two basic kinds of elementary particles that make up matter in the fascinating world of particle physics. Let's look at the qualities and importance of these particles.

Quarks:

Quarks are fundamental particles with fractional electric charges that serve as the foundation for protons and neutrons, which comprise the atomic nucleus. Quarks are classified into six distinct types: up, down, charm, strange, top, and bottom. Each flavour of quark has a unique mix of charge and mass.

Up and down quarks: The up quark contains a charge of +2/3 (in elementary charge units) and is lighter than the down quark, which carries a charge of -1/3. These two types of quarks are the most common in ordinary matter and are responsible for atomic nuclei's stability.

Charm and strange quarks: The charges of charm and strange quarks are +2/3 and -1/3, respectively. These quark types are more heavy than up and down quarks and are frequently created in high-energy particle collisions or live in extreme environments like neutron stars.

Top and Bottom Quarks: The top quark, commonly known as the "truth" quark, is the heaviest known elementary particle, with a charge of $+2/3$. The bottom quark is also relatively heavy, with a charge of $-1/3$. These quarks are often created in high-energy particle accelerator operations and help us understand particle physics.

Leptons:

Leptons, like quarks, are a type of elementary particle that does not participate in the strong nuclear interaction. Leptons are particles that we are familiar with, such as electrons, muons, and neutrinos.

Electrons: Electrons are negatively charged particles that orbit around the atomic nucleus and participate in chemical reactions. In comparison to quarks, they have a very little mass and are stable in most situations.

Muons: Muons are identical to electrons, but much heavier. After a small time period of existence, they disintegrate into electrons and other particles. Muons can be created in particle accelerators or by cosmic ray interactions in the Earth's atmosphere.

Neutrinos: Neutrinos are incredibly light, neutral particles that interact with matter only weakly. They are categorised as electron neutrinos, muon neutrinos, and tau neutrinos. Neutrinos are particles that are constantly created in numerous astrophysical processes and can reveal important information about the universe's features.

Understanding the characteristics of quarks and leptons is crucial for understanding matter's composition and particle behaviour at the most fundamental level. The Standard Model of particle physics is built on these constituent particles and their interactions, and it provides a framework for understanding the fundamental forces and the nature of the world.

Electromagnetic Force

According to the Standard Model of particle physics, the electromagnetic force as we discussed earlier also is one of the four fundamental forces in existence. It is in charge of interactions involving electrically charged particles and plays a critical role in our understanding of electromagnetism and charged particle behaviour.

The exchange of particles known as photons is at the heart of the electromagnetic force. Photons are massless particles that carry electromagnetic force. They mediate the interaction of charged particles, allowing electromagnetic energy to be transmitted and electromagnetic fields to be generated.

Electrons and protons, for example, interact with each other via exchanging photons. This interaction produces two essential features of electromagnetic force: attraction and repulsion. Like charges repel each other, whereas opposite charges attract each other. This behaviour can be seen in everyday phenomena such as two magnets repelling or attracting each other or garments clinging together after being statically charged.

The electromagnetic force governs the behaviour of charged particles in the presence of electric and magnetic fields, in addition to interactions between charged particles. Electric charges generate electric fields, which apply forces on other charges in their close vicinity. Moving charges or changing electric fields, on the other hand, generate magnetic fields, which can exert forces on charged particles.

Electromagnetic waves are created when electric and magnetic fields interact, and these waves include visible light, radio waves, microwaves, X-rays, and gamma rays. These waves travel through space, taking energy and information with them. The wavelengths and frequencies of electromagnetic waves govern their properties such as colour (in the case of visible light) or their capacity to penetrate particular materials (in the case of X-rays).

The electromagnetic force is important in many parts of our daily life, from the operation of electrical equipment to the behaviour of light and the transmission of information via electromagnetic waves. It is also the foundation of electric circuits, electromagnetism, and optics. Understanding the electromagnetic force is critical for understanding a wide range of phenomena and is critical in fields such as electrical engineering, telecommunications, and optics.

Strong Nuclear Force

The strong nuclear force, commonly known as the strong interaction or the strong force, is one of the four fundamental forces in particle physics as described by the Standard Model. It is in charge of holding atomic nuclei together and binding quarks together to produce protons and neutrons, the building blocks of atomic nuclei.

The particles known as gluons are at the centre of the strong nuclear force. Gluons are massless particles with a colour charge connected with the strong force. Gluons mediate the strong force by swapping themselves between quarks, just like photons do for the electromagnetic force.

Quarks, as previously discussed, are elementary particles that come in a variety of flavours, including up, down, charm, strange, top, and bottom. The strong force interacts with quarks, allowing them to be linked together to create composite particles like protons and neutrons. The strong force is distinct in that its strength increases with distance, implying that the interaction between two quarks grows stronger as they are drawn apart. Asymptotic freedom is a result of the underlying theory of quantum chromodynamics (QCD), which describes the strong force.

The phenomenon of confinement is an important feature of powerful forces. Quarks are never encountered as isolated free particles, but only in bound states. The strong force becomes stronger as quarks are separated, making it favourable for the formation of new quark-antiquark pairs from the vacuum. This prevents individual quarks from being isolated and keeps them bound within composite particles.

The strong force is crucial in understanding matter's stability and characteristics. It supplies the binding energy that holds atomic nuclei together, allowing stable atoms to exist. Furthermore, the strong force is involved in nuclear reactions like fusion and fission, which release significant amounts of energy and are used in applications like nuclear power.

The study of the strong nuclear force and its properties is critical in particle and nuclear physics. Understanding the strong force is critical for interpreting the complexity of atomic nuclei, subatomic particle behaviour, and the nature of matter. We get deeper insights into the fundamental structure of the universe by exploring the strong force.

Weak Nuclear Force

The weak nuclear force, commonly known as the weak interaction, is one of the four fundamental forces described by particle physics' Standard Model. It is in charge of a variety of subatomic particle processes, including as beta decay and neutrino interactions. The weak force differs from the other fundamental forces in that it acts over a relatively short distance and participates in processes that change the fundamental properties of particles.

As previously mentioned The W and Z bosons are the weak force's carrier particles. There are two forms of W bosons: W+ and W-, which carry positive and negative electric charges, respectively. In contrast, the Z boson is electrically neutral. These bosons are in charge of mediating weak particle interactions.

Particles in weak interactions can change character or change into distinct particles. For example, beta decay involves the conversion of a neutron into a proton, which results in the emission of an electron (or positron) and an electron antineutrino (or neutrino). The exchange of a W boson mediates this process. Neutrinos can also interact with other particles weakly, such as via scattering neutrinos off electrons or nucleons.

Weak isospin is an important concept in understanding weak interactions. Weak isospin is a quantum characteristic of particles that governs their behaviour in the presence of a weak force. The weak force interacts differentially amongst particles with varying weak isospin values. This feature is identical to electric charge, except that weak isospin governs weak interactions rather than electromagnetic.

Neutrino oscillation is a surprising finding connected to the weak force. Neutrinos are electrically neutral particles with extremely small masses. They are classified into three types: electron neutrinos, muon neutrinos, and tau neutrinos. Neutrino oscillation is the phenomenon in which neutrinos of one flavour spontaneously convert into neutrinos of another flavour as they travel through space. This result implies that neutrinos have non-zero masses, which the Standard Model did not foresee. Neutrino oscillation experiments have offered solid evidence for neutrino masses' existence and have opened up new paths for exploring neutrino characteristics.

Understanding the behaviour of subatomic particles and the processes that occur within atomic nuclei requires an understanding of the weak nuclear force. It is important for matter stability as well as the kinetics of nuclear reactions and particle decays. The study of the weak force has yielded important information into the nature of fundamental particles and their interactions, which has helped us realise the fundamental forces that govern the universe.

Gravitational Force

Although it is not specifically included in the Standard Model of particle physics, gravitational force is one of nature's fundamental forces. Gravity, as described by general relativity theory, plays an important role in the behaviour of enormous objects, the dynamics of celestial bodies, and the construction of the cosmos.

Gravitational force is created by the bending of spacetime induced by mass and energy. Massive objects, according to general relativity, distort the fabric of spacetime around them, creating a gravitational field. Other things, whether particles or celestial bodies, then move along the curvature-determined trajectories.

Gravity influences the mobility and interactions of fundamental particles. While the gravitational interaction between elementary particles is exceedingly weak in comparison to other fundamental forces like electromagnetic or strong nuclear forces, it becomes substantial on cosmological scales or in the presence of enormously large objects like black holes. For example, gravitational force effects particle behaviour in the vicinity of big astronomical objects or during cosmic processes such as star collapse or galaxy creation.

Gravity's universality is one of its most striking characteristics. It has an effect on all particles, regardless of mass or charge. Gravity, unlike the other Standard Model forces, does not discriminate depending on particle characteristics. It affects all types of matter and energy, including elementary particles, atoms, planets, and stars, as well as light itself.

General relativity provides a comprehensive framework for understanding gravity, outlining how it warps spacetime and affects particle motion. Numerous tests and observations have thoroughly verified and validated it, including the bending of starlight by big objects, the redshift of light in gravitational fields, and exact measurements of planetary orbits.

The study of Gravity is critical for understanding the behaviour of huge objects and the large-scale structure of the universe. It is crucial in cosmology, astrophysics, and the study of black holes, gravitational waves, and the evolution of the cosmos itself. While the gravitational force is not explicitly unified with the other forces in the Standard Model, ongoing research aims to reconcile general relativity with quantum mechanics, leading to the development of theories such as quantum gravity.

In summary, the gravitational force is an important part of our understanding of the cosmos. It controls particle motion, shapes the cosmos on both tiny and big sizes, and remains an important topic of research as we seek to understand the complicated relationship between gravity, quantum mechanics, and the underlying fabric of spacetime.

We get an improved understanding of the underlying building blocks of matter and the interactions that create the physical world by studying elementary particles and fundamental forces. Although the Standard Model provides a thorough framework for understanding these particles and forces, it is crucial to emphasise that there are still unresolved problems and ongoing research to discover the mysteries of the universe beyond the Standard Model.

4.3 Particle Accelerators and Colliders

Introduction

Now we are going to enter the fascinating world of particle accelerators and colliders, massive equipment designed to examine nature's fundamental particles and forces. These technological wonders are critical in furthering our understanding of the cosmos because they allow scientists to investigate matter at the lowest scales and replicate the circumstances that existed seconds after the Big Bang.

Introduction to Particle Accelerators

Particle accelerators are at the centre of scientific research, allowing us to explore the fundamental building blocks of matter and solve cosmic mysteries. These extraordinary gadgets are strong tools for accelerating charged particles such as protons or electrons to near-light speeds. By doing so, they enable scientists to explore the fundamental features of particles and their interactions.

Particles are accelerated in a particle accelerator by carefully applying electric fields, magnetic fields, or a combination of both. The electric fields generate an electric potential that accelerates the charged particles, while the magnetic fields control their direction, ensuring that they go along a predetermined path within the accelerator.

When the particles reach the desired energy level, they are steered towards a target or collide with other accelerated particles. These high-energy collisions produce strong bursts of energy as well as a number of new particles. Scientists can learn about fundamental forces, particles, and their properties by analysing the debris and particle characteristics produced in these collisions.

Particle accelerators exist in a variety of shapes and sizes, each with its own set of scientific purposes and energy requirements. Linear accelerators (linacs) use straight lines to accelerate particles, whereas circular accelerators (synchrotrons) use magnetic fields to bend the particles into a circular route. Colliders, which are more intricate designs, bring together beams of particles travelling in opposite directions to achieve head-on collisions.

Particle accelerators have enabled outstanding developments in the field of particle physics. These machines have helped to prove the existence of new particles, such as the Higgs boson, and to explore phenomena such as quark-gluon plasma and neutrino oscillations. They have contributed experimental data that has changed our understanding of the fundamental particles and forces that control our cosmos.

Particle accelerators have practical uses in sectors such as medicine, material science, and industry, in addition to fundamental research. They are used, for example, to generate powerful radiation beams for cancer therapy, to research the characteristics of materials under extreme conditions, and to generate high-energy particle beams for industrial processes.

Particle accelerators will play an important role in figuring out the mysteries of the universe and unlocking new worlds of discovery as we continue to push the boundaries of scientific knowledge. We should expect even more amazing discoveries and scientific breakthroughs in the future as accelerator technologies improve and larger and more powerful devices are built.

Types of Particle Accelerators

Particle accelerators are available in a variety of shapes and sizes, each adapted to specific scientific goals and energy requirements. Let's look at some of the most common types of particle accelerators used in scientific research:

Linear Accelerators (Linacs): Linear accelerators, also known as linacs, use alternating electric fields to accelerate particles in a straight line. These accelerators are made up of long, evacuated tubes in which electric fields drive particles ahead. Linacs are frequently employed as injectors for larger accelerators or as standalone units for specialised study. They can achieve high energies, but their maximum energy is limited due to practical limitations.

Circular Accelerators (Synchrotrons): Magnetic fields are used to bend particles into a circular path in circular accelerators, also known as synchrotrons. They have a ring-shaped structure with powerful magnetic fields that accelerate and direct particles. Electric fields are utilised to accelerate particles as they circulate in the synchrotron. Synchrotrons are more powerful than linacs and are employed in a wide range of scientific disciplines, including particle physics, nuclear physics, and materials research.

Cyclotrons: Cyclotrons are a form of circular accelerator that uses an electric and magnetic field combination to accelerate particles in a spiral route. They work on the resonance principle, in which particles gather energy as they pass through a sequence of increasing gaps between electrodes. Cyclotrons are typically small and utilised for low-energy applications such as the production of medical isotopes for imaging and cancer therapy.

Storage Rings: Storage rings are circular accelerators that do not aim to improve particle energy but rather to keep it constant. Particles in storage rings revolve in a closed loop, guided by magnetic fields. These rings are used for a variety of applications, including synchrotron light sources that generate powerful X-ray and other electromagnetic radiation, as well as researching the behaviour of charged particles over long periods of time.

Colliders: Colliders are a unique kind of circular accelerator in which two streams of particles collide. The beams circulate in opposite directions and collide head-on at designated interaction sites. Colliders enable scientists to explore particle properties and interactions by analysing the debris produced in collisions. Colliders include the Large Hadron Collider (LHC) at CERN, which was essential in the discovery of the Higgs boson.

These were only a few examples of particle accelerators that are utilised in scientific study. Each accelerator design has distinct advantages and is customised to specific energy ranges and experimental requirements. Accelerator advances continue to push the boundaries of what is possible, allowing scientists to explore new frontiers in particle physics, materials research, and a variety of other scientific disciplines.

Large Hadron Collider (LHC)

The Large Hadron Collider (LHC) is a groundbreaking particle accelerator and collider located at CERN near Geneva, Switzerland. It is the world's largest and most powerful particle collider, capable of accelerating protons or heavy ions to energies never before attained in a laboratory setting. The

LHC is critical to furthering our understanding of particle physics and investigating the fundamental aspects of matter and the universe.

Design and Construction:

The LHC's design and construction were ambitious, involving thousands of scientists, engineers, and technicians from around the world. The accelerator is a 27-kilometer (17-mile) subterranean ring that is buried beneath the French-Swiss border. It is a superconducting machine, which means it runs at extremely low temperatures by cooling the magnets using superfluid helium to -271.3 degrees Celsius (-456.3 degrees Fahrenheit).

The LHC is made up of two parallel beam pipes that alternately circulate protons or heavy ions. A complicated system of RF cavities and superconducting magnets is used to accelerate these beams. The magnets are in charge of directing and concentrating the particle beams, while the radiofrequency cavities give the energy boost required to raise particle velocity.

The LHC works by progressively increasing the energy of the particle beams until they reach their maximum intended energy. Proton beams can be accelerated to energies of up to 6.5 TeV, whereas heavier ions, such as lead ions, can reach energy of many teraelectron volts per nucleon. Once the beams have attained the necessary energy level, they are directed to collide at four major experimental sites located along the LHC ring.

Large detectors, such as ATLAS and CMS, are placed at these collision locations. These detectors are intended to capture and analyse the particles generated by collisions. The LHC has extraordinary collision energy, allowing scientists to recreate circumstances that existed fractions of a second after the Big Bang and explore the properties of particles that existed in the early cosmos.

Scientific Discoveries:

The LHC has played a role in a number of ground-breaking scientific discoveries. The ATLAS and CMS experiments at the LHC announced the discovery of the Higgs boson, a particle important in explaining the origin of mass, in 2012. This finding established the existence of the Higgs field, which pervades the universe and determines particle masses.

In addition to the Higgs boson, the LHC has enabled remarkable precision in the investigation of other fundamental particles, including as the top quark and the W and Z bosons. It has also contributed to our understanding of neutrino characteristics, the quest for dark matter particles, and the discovery of new physics outside the Standard Model.

Future Upgrades: The LHC continues to push the boundaries of scientific investigation, and plans for future upgrades are in the works. The High-Luminosity LHC (HL-LHC) project seeks to greatly boost the collider's luminosity, allowing for more collisions and greater precision observations. Scientists will be able to examine unusual processes and occurrences with higher sensitivity due to this update.

The LHC and its continuous research contribute significantly to our understanding of the fundamental nature of matter and the universe. Its discoveries and advances not only impact our understanding of particle physics, but they also have far-reaching implications for our understanding of the universe and the fundamental principles that govern it.

Particle accelerators and colliders, in general, have altered our understanding of the fundamental building components of matter and the forces that control them. These amazing devices continue to propel scientific progress and impact our understanding of the universe, opening up new horizons for exploration and discovery.

4.4 The Higgs Boson and the Origins of Mass

Introduction

Now we are going to investigate the mysterious Higgs boson and its role in explaining the origins of mass. The discovery of the Higgs boson at the Large Hadron Collider (LHC) in 2012 was a watershed moment in particle physics, proving the existence of a particle long thought to be responsible for the mass of other particles in the universe.

Introduction to Higgs Field

The Higgs field is an important concept in current particle physics that is important in understanding the origins of mass in the cosmos. The Higgs field, proposed by scientist Peter Higgs and others in the 1960s, is an invisible and widespread field that permeates all of space.

The Higgs field, like other fundamental fields such as the electromagnetic field, is considered a fundamental field in the framework of particle physics' Standard Model. Unlike the electromagnetic field, which produces electromagnetism's force, the Higgs field is responsible for endowing particles with mass.

The Higgs field is distinguished by its unique quality known as non-zero vacuum expectation value. In simple language, this indicates that the Higgs field has a non-zero value even in empty space, filling the cosmos with its existence. Particles going through the Higgs field encounter drag or resistance, similar to things moving through a viscous medium.

This interaction with the Higgs field is responsible for particle mass. The stronger a particle interacts with the Higgs field, the larger it grows. Particles that interact weakly with the Higgs field, such as photons, remain massless, but particles that engage more strongly with the Higgs field, such as quarks and electrons, gain mass through their interactions with the Higgs field.

The Higgs field is considered to exist in all of space, interacting with particles all around the universe. Its existence and qualities have far-reaching impacts on our understanding of matter's fundamental nature. The Higgs field has been a key focus of experimental efforts, culminating in the discovery of the Higgs boson at the Large Hadron Collider (LHC) in 2012.

The Higgs field and its related particle, the Higgs boson, give a method for understanding why and how particles gain mass. The discovery of the Higgs field has transformed our knowledge of the fundamental building blocks of the cosmos, opening us new possibilities for investigating physics beyond the Standard Model.

Mechanism of Mass Generation

The Higgs mechanism is the process by which particles acquire mass by their interaction with the Higgs field. Particles develop mass as a result of their interactions with the Higgs field, according to this mechanism.

The Higgs field covers the entire universe, filling everything with its existence. Particles gain their mass through this field. Consider a simplistic analogy to better understand the mechanism of mass formation.

Consider a pool of water to represent the Higgs field. Consider an item travelling through this pool. The object encounters resistance or drag from the water as it moves. This resistance slows the thing down and necessitates an energy input to overcome it.

Similarly, particles passing through the Higgs field encounter resistance or drag, which causes them to gain mass. The interaction between the particles and the Higgs field causes this drag. The stronger the interaction, the larger the drag, and thus the particle's mass.

Photons and other particles that interact weakly with the Higgs field experience little resistance and remain massless. Particles that interact more strongly with the Higgs field, such as quarks and electrons, experience severe drag and gain mass as a result of their interactions.

The Higgs mechanism explains the origin of elementary particle mass in the Standard Model. It provides a way for particles to acquire mass while maintaining symmetry and gauge invariance.

As mentioned earlier the discovery of the Higgs boson at the Large Hadron Collider (LHC) in 2012 was a crucial step towards establishing the Higgs field's existence and validating the mechanism of mass creation. The Higgs boson is a Higgs field excitation, and its discovery offered direct evidence of the Higgs process in action.

Further research and experiments are being conducted to examine the properties and interactions of the Higgs boson, offering information on the complex nature of mass creation in the cosmos. Understanding the Higgs process and its function in mass creation is an important achievement in particle physics, adding to our understanding of the fundamental forces and particles that control the world.

The Discovery of Higgs Boson

The discovery of the Higgs boson was an important turning point in the history of particle physics. It was the outcome of intensive experimental efforts at CERN, the European Organisation for Nuclear Research's Large Hadron Collider (LHC).

The world's most powerful particle accelerator, the LHC, was designed to collide protons at extremely high energies. These high-energy collisions created conditions similar to those only minutes after the Big Bang, allowing scientists to investigate the universe's core building blocks.

The necessity to confirm the existence of the Higgs field, which is responsible for particle mass, drove the quest for the Higgs boson. The Higgs boson is a particle linked with Higgs field fluctuations. Its detection would provide direct confirmation of the Higgs mechanism and validate our understanding of the universe's mass creation.

The LHC experiments involved smashing protons at energies of 7 TeV and eventually 13 TeV, resulting in massive amounts of data. The data was obtained using advanced detectors like as the ATLAS and CMS detectors, which were designed to catch and analyse particle collision products.

The discovery of the Higgs boson involved the examination of massive amounts of data in search of evidence indicating the presence of this elusive particle. Scientists had to go through an enormous number of collision events to find ones that might have contained Higgs bosons.

The ATLAS and CMS groups at the LHC announced the finding of a new particle consistent with the Higgs boson on July 4, 2012. The discovery was based on an examination of data gathered throughout the studies. The measured particle's properties, particularly its decay modes and mass, were compatible with those predicted by the Standard Model.

The finding of the Higgs boson was a significant scientific accomplishment since it confirmed the existence of the Higgs field and validated the mechanism by which particles acquire mass. It was the result of decades of theoretical and experimental research by physicists all around the world.

This discovery is significant for reasons other than the confirmation of a theoretical concept. It clarifies the fundamental nature of the cosmos and the fundamental forces that control it. The discovery of the Higgs boson has opened up new study routes, allowing scientists to study the origins of mass, the nature of the Higgs field, and even discover physics outside the Standard Model.

Since its discovery, numerous studies and experiments have been done to investigate the Higgs boson's characteristics and interactions. These studies seek to expand our understanding of the underlying forces and particles that shape the universe, paving the path for new discoveries and advances in our understanding of the universe.

Properties and Interactions of Higgs Boson

The Higgs boson, as found at the Large Hadron Collider (LHC), has a number of significant features. The mass of the Higgs boson is one of its most important features. The accurate measurement of the Higgs boson's mass at the LHC, which was calculated to be around 125 gigaelectronvolts (GeV), verified the existence of this elusive particle.

The Higgs boson has a distinct spin in addition to its mass. Spin is a characteristic of elementary particles that defines their angular momentum. The Higgs boson is a scalar particle, meaning it has no spin. This distinguishes it from other particles with spin values of 1/2, such as quarks and leptons.

The Higgs boson's primary purpose is to interact with other particles and provide them with mass. Particles gain mass by interactions with the Higgs field, according to the Higgs mechanism. The carrier particle in these interactions is the Higgs boson. When particles move through the Higgs field, they suffer a drag-like interaction, which causes them to gain mass.

The strength of the interaction between the Higgs boson and other particles is determined by the couplings of the particles to the Higgs field. The Higgs boson interacts most strongly with heavier particles, such as the W and Z bosons. These gauge bosons, which carry the weak nuclear force, develop mass by interacting with the Higgs field and exchanging Higgs bosons.

The Higgs boson interacts with other basic particles, such as quarks, leptons, and photons, though at a weaker level. These interactions are critical to comprehending the origins of mass in the universe. Scientists can research the underlying mechanisms that influence the behaviour of matter by studying the Higgs boson's interactions with other particles.

The discovery and characterization of the Higgs boson has opened up new paths for particle physics research. Scientists continue to study its features and interactions in greater depth in order to acquire a better understanding of the fundamental nature of the universe. Current LHC experiments and future collider efforts aim to investigate the Higgs boson's interactions with greater accuracy, understanding information on the mysteries of mass, the nature of the Higgs field, and the potential presence of novel physics outside the Standard Model.

Understanding the Higgs boson's characteristics and interactions is critical not just for developing particle physics but also for solving the mysteries of the universe's fundamental forces and the origins of mass. The study of the Higgs boson provides important insights into the nature of matter, providing a greater awareness of the universe's building elements.

Implications and Significance

The discovery of the Higgs boson has significant implications for our understanding of the universe and its laws. Here, we look at the broader implications of this discovery and its implications for physics and cosmology.

Confirmation of the Standard Model: The existence of the Higgs boson provides critical experimental validation of the Higgs mechanism and the Standard Model of particle physics. It lends support to the theoretical framework that describes fundamental particles and their interactions. The discovery closes a fundamental gap in our knowledge of elementary particles and their characteristics in the universe.

Origin of Particle's Mass: The Higgs boson's involvement in giving mass to other particles via interactions with the Higgs field helps to explain one of particle physics' fundamental mysteries: the genesis of mass. The discovery of the Higgs boson provides insights into the fundamental basis of matter and the structure of the cosmos by revealing how particles gain mass.

Early Universe and Cosmology: The Higgs field, with which the Higgs boson is related, proposes that during the early moments of the universe's origin, a field pervaded all of space, interacting with particles and endowing them with mass. The implications for cosmology and our understanding of the early cosmos are significant. It clarifies the conditions and processes that occurred throughout the early stages of cosmic history.

Beyond the Standard Model: While the discovery of the Higgs boson validates the Standard Model's predictions, it also raises new problems. The Standard Model is widely acknowledged as an insufficient theory because it excludes gravity and fails to explain other phenomena such as dark matter and dark energy. The study of the Higgs boson and its properties provides vital clues that could drive the development of new theories, such as supersymmetry or extra dimensions, aimed at extending the Standard Model and answering these unresolved issues.

Technological Applications: Research and technology generated in the goal of identifying and researching the Higgs boson have resulted in various technological improvements. Particle accelerators and detectors, such as the LHC, have pushed scientific instrumentation and engineering to new heights. These advances have far-reaching applications outside of particle physics, including medical imaging and cancer treatment, as well as materials science and computation.

To summarise, the discovery of the Higgs boson has helped us better comprehend the fundamental particles and forces that govern our universe. It has provided experimental confirmation of the Standard Model, shed light on the origin of particle mass, and revealed information about the early cosmos. Furthermore, it has opened up new paths for scientific research and technological growth, while also inspiring and fascinating scientists and the general public.

Beyond the Standard Model

The discovery of the Higgs boson not only validated the Standard Model's predictions, but also exposed its limitations. While the Standard Model successfully depicts fundamental particles and their interactions, it fails to account for gravity and does not explain several known phenomena in the universe. We will look at some of the ongoing studies and theories that try to go beyond the Standard Model of particle physics.

Supersymmetry (SUSY): Supersymmetry is one of the most extensively researched hypotheses beyond the Standard Model. It suggests the existence of superpartners for every known particle, hence providing a symmetry between bosons and fermions. Supersymmetry could solve the hierarchy problem and provide a contender for dark matter. Although the LHC experiments are yet to find evidence of supersymmetry, study in this subject is ongoing, with researchers investigating various supersymmetric theories and their consequences.

Grand Unified Theories (GUTs): GUTs strive to combine the electromagnetic, weak, and strong forces into a single framework. According to these views, at high energies, the three forces are diverse manifestations of a single undivided force. GUTs may help to explain observed trends in particle masses and charge quantization. They also anticipate the existence of new particles and events, which may be investigated in future tests.

Extra Dimensions: For decades, physicists have been fascinated by the concept of more spatial dimensions beyond the three we experience. Theories such as the Kaluza-Klein and string theories propose the existence of additional compactified dimensions, which could explain the particle mass hierarchy and the nature of gravity. Although experimental verification of extra dimensions remains difficult, ongoing research focuses on their possible implications for particle physics and cosmology.

Dark Matter and Dark Energy: The presence of dark matter, as inferred from its gravitational effects, and dark energy, which causes the universe's accelerating expansion, are two of modern physics' most significant mysteries. Many ideas outside the Standard Model propose additional particles as potential candidates for dark matter, such as weakly interacting massive particles (WIMPs), axions, or sterile neutrinos. The study of the nature of dark matter and dark energy is still ongoing.

Neutrino Physics: Neutrinos, which were previously thought to be massless in the Standard Model, have been experimentally observed to have minuscule but non-zero masses. This discovery prompted the development of ideas that go beyond the Standard Model to explain the observed neutrino oscillations and account for their masses. To understand the nature of neutrinos and their role in the cosmos, extensions to the Standard Model, such as seesaw processes or sterile neutrinos, have been proposed.

These are only a few of the many theories and research directions that have been investigated in order to expand our understanding of particle physics beyond the Standard Model. Scientists continue to hunt for evidence of new particles, occurrences, and fundamental principles that can improve our understanding of the cosmos and its underlying physics as experimental techniques evolve and new data is obtained. The search for a more comprehensive theory that unites all known forces and explains the cosmos' mysteries continues to be a fascinating and active subject of study.

The discovery of the Higgs boson, as well as our knowledge of the Higgs field's function in mass creation, are key milestones in our investigation of the fundamental nature of the cosmos. The discovery of the Higgs boson has deepened our understanding of the forces and particles that shape our world, setting the path for future discoveries and advances in particle physics.

Chapter 5

Seeking The Theory of Everything

5.1 Supersymmetry and Superpartners

Introduction

This section explores the concept of supersymmetry (SUSY), a proposed extension to particle physics' Standard Model. Supersymmetry implies the existence of a fundamental symmetry between bosons and fermions, resulting in the creation of a new class of particles known as superpartners or sparticles. We investigate the implications of supersymmetry, its theoretical framework, and the potential consequences of its discovery.

Introduction to Supersymmetry

Supersymmetry (SUSY) is a theoretical framework that proposes a significant extension to the Standard Model of particle physics. It introduces the concept of a novel symmetry between particles with differing spins, implying that each known basic particle has superpartners. Except for a variation in spin statistics, these superpartners have the same quantum numbers as their Standard Model counterparts.

The aim of supersymmetry is to offer a deeper level of symmetry in nature by connecting fermions (particles with half-integer spin) to bosons (particles with integer spin). This symmetry introduces a new type of symmetry transformation called supersymmetry transformations, which connects the two classes of particles. The supersymmetry transformations result in a rich mathematical framework that unites particles with varying spin characteristics.

One of the primary motivations for supersymmetry is its ability to address the particle physics hierarchy problem. The hierarchy problem refers to the difference between the weak and gravitational forces. The weak force is approximately 10^{32} times stronger than gravity, and the Standard Model cannot account for this enormous difference. Supersymmetry solves this difficulty by cancelling out some quantum corrections to the Higgs boson mass and therefore stabilising the electroweak scale.

Furthermore, supersymmetry is a possible candidate for dark matter, which is thought to account for a considerable percentage of the universe's mass. The LSP (lightest supersymmetric particle) is projected to be stable and electrically neutral, making it a good candidate for dark matter. The LSP, if discovered, could help explain the mysterious nature of dark matter and provide a better understanding of the universe's appearance.

While supersymmetry is a beautiful and intriguing concept, its existence has yet to be proved by experimental evidence. Particle accelerators, such as the Large Hadron Collider (LHC), have aided in the search for superpartners, although no direct evidence of supersymmetry has been discovered so far. Nonetheless, supersymmetry research continues to drive theoretical and experimental research in particle physics, providing opportunities for understanding the underlying nature of the universe.

Theoretical Framework

Supersymmetry, as a theoretical framework, is based on the idea of expanding space and time symmetries to include an additional set of transformations that connect particles with different spins. These new transformations, known as supersymmetry transformations, add a new level of symmetry to nature.

Each particle in the Standard Model is coupled with a superpartner, known as a sparticle, which differs in spin by half a unit in supersymmetry. Fermions having half-integer spin link up with bosonic superpartners with integer spin, and vice versa. For example, the electron, a fermion with spin 1/2, is connected with the selectron, a bosonic superpartner with spin 0.

Supersymmetry is mathematically characterised by a collection of symmetry operators known as supercharges that act on particle states. These supercharges are responsible for the supersymmetry transformations that link fermionic and bosonic states. The supersymmetry algebra defines the supercharge commutation and anticommutation relations and serves as the foundation for supersymmetric theories.

The concept of superfields is introduced in order to develop supersymmetric theories. Superfields are mathematical constructions that include particles' bosonic and fermionic degrees of freedom. They encode in a coherent way the properties and interactions of particles and their superpartners. Superfields enable the development of supersymmetric Lagrangians, which characterise the dynamics of supersymmetric theories.

A cancellation of quadratic divergences is a notable property of supersymmetric Lagrangians. These divergences, caused by quantum corrections to particle masses, call the theory's naturalness into question. However, in supersymmetry, the contributions of the superpartners cancel out these divergences, resulting in a more elegant and theoretically appealing framework.

Theoretical physicists have created a number of supersymmetric theories, including the Minimal Supersymmetric Standard Model (MSSM), which extends the Standard Model by incorporating a set of superpartners for each known particle. These theories combine supersymmetry principles and attempt to explain phenomena that the Standard Model alone cannot explain, such as the hierarchy problem and the nature of dark matter.

Although supersymmetry provides elegant answers to these fundamental concerns, experimental confirmation of supersymmetric particles is required to substantiate its predictions. Particle accelerators, such as the LHC, have been looking for evidence of sparticles, but no direct observations have been made as of yet. Nonetheless, the theoretical framework of supersymmetry remains an active area of research, sparking fresh ideas and opening up new paths for investigating the underlying basis of our world.

Superpartners and their Interactions

In the Standard Model, supersymmetry predicts the existence of superpartners for each particle. These superpartners have different spins and are predicted to have properties similar to their respective particles. Let's look at some of the expected superpartners and how they interact.

Selectron: The superpartner of the electron is the selectron. It should have the same electric charge as an electron but a different spin. The selectron, like the electron, can interact with other particles via electromagnetic interactions.

Squark: Squarks are quarks' superpartners. They are classified according to the flavour of the quark (up, down, charm, strange, top, and bottom). Squarks, like quarks, have colour charge and participate in strong interactions mediated by gluons.

Gauginos: Gauginos are the superpartners of gauge bosons, which are fundamental force carrier particles. The photon's superpartner (carrier of the electromagnetic force) is known as the photino, whereas the W and Z bosons' superpartners (carriers of the weak force) are known as winos and zinos, respectively. Gauginos are believed to exhibit properties similar to their counterpart gauge bosons, such as charge and spin.

Higgsinos: Higgsinos are the Higgs boson's superpartners. Higgsinos are expected to interact with other particles via the Higgs field, in the same way as the Higgs boson interacts with particles to give them mass.

The Standard Model describes the fundamental forces that govern superpartner interactions with other particles. For example, photon exchange mediates electromagnetic interactions between superpartners and charged particles. Gluons mediate strong interactions involving squarks and quarks, while the W and Z bosons mediate weak interactions involving gauginos and other particles.

Understanding the properties and interactions of superpartners is critical for testing supersymmetry predictions and investigating their implications for particle physics. High-energy particle colliders, like as the LHC, are used to hunt for these superpartners in the goal of discovering evidence for supersymmetry and learning more about the nature of our universe's fundamental particles and forces.

Experimental Signatures

The investigation of experimental evidence that could suggest the presence of superpartners in high-energy collisions is part of the hunt for supersymmetry. Particle colliders like the Large Hadron Collider (LHC) are critical to these efforts. We cover some of the experimental indications of supersymmetry as well as the techniques utilised to find superpartners.

Missing Transverse Energy (MET): The occurrence of considerable missing transverse energy is a distinguishing feature of supersymmetry. When superpartners, such as neutrinos, exit the detector without leaving a detectable trace, this happens. The conservation of momentum in the transverse plane can be used to infer the presence of unseen particles and to confirm the presence of supersymmetry.

Same-Sign Dileptons: Certain decay processes in some supersymmetric models can yield pairs of same-sign charged leptons (electrons or muons). This is extremely implausible in the Standard Model, but it can be a strong indicator of supersymmetry. Detecting an excess of same-sign dilepton events above the background predictions could be evidence of superpartner formation and decay.

Jets and Missing Transverse Energy: The generation of superpartners in several supersymmetric models can result in the emission of high-energy jets (particle sprays) in conjunction with missing transverse energy. Jets are particle clusters generated when quarks or gluons fracture and hadronize. The presence of an excess of high-energy jets in the presence of a substantial missing transverse energy can be an evidence of supersymmetry.

Resonance Signatures: Supersymmetry can also appear as resonance peaks in the invariant mass distribution of some particles. The decay of a superpartner, for example, may provide a distinct peak in the dijet or dilepton invariant mass spectrum. Observing resonance signals that are not explained by known particles would be significant proof for supersymmetry.

These are only a few of the experimental signals sought by researchers in their search for superpartners and the presence of supersymmetry. The large amount of data produced at particle colliders is analysed using advanced detectors, sophisticated data analysis tools, and strong computer algorithms, looking for departures from the predicted background and discovering supersymmetry signals.

Although experimental data have yet to prove supersymmetry, current searches at the LHC and future collider experiments continue to push the boundaries of our understanding of particle physics and explore the possibility of unusual events beyond the Standard Model.

Constraints and Current Status

For several decades, researchers have been looking for supersymmetry. While the theoretical framework of supersymmetry provides elegant answers to several unresolved puzzles in particle physics, the lack of clear experimental evidence for superpartners has limited the parameter space of supersymmetric models.

LHC Constraints: By colliding particles at unprecedented energies, the Large Hadron Collider (LHC) has played a critical role in proving supersymmetric ideas. Despite extensive searches, no major departures from the Standard Model predictions that can be attributed to the formation and decay of superpartners have been detected. This has resulted in the placement of limits on masses and superpartner couplings, thereby ruling out certain areas of parameter space.

Naturalness and Fine-Tuning: The idea of naturalness, which indicates that the fundamental parameters of a theory should not require extreme fine-tuning to produce the known aspects of the cosmos, is one of the inspirations behind supersymmetry. The lack of superpartner findings at the LHC has prompted concerns about the naturalness of supersymmetry, as some models may require extensive fine-tuning to match experimental observations.

Theoretical considerations: Aside from experimental restrictions, theoretical concerns influence the current state of supersymmetry. The lack of direct evidence for superpartners, paired with the growing energy scale at which new physics may appear, has sparked debate regarding alternate scenarios and Standard Model extensions that go beyond traditional supersymmetry.

Supersymmetry and Dark Matter: Supersymmetry provides a convincing candidate for dark matter, such as the lightest neutralino, which is stable and weakly interacting. The lack of direct evidence for superpartners at colliders has implications for dark matter searches and the relationship between supersymmetry and dark matter abundance in the cosmos.

While the lack of conclusive evidence for supersymmetry has hampered the theory, it is crucial to remember that several parameter regions remain plausible, and supersymmetry remains an active topic of research. Ongoing LHC experiments, as well as future collider experiments and complementary researches, such as direct and indirect dark matter detection experiments, will deepen our understanding of the parameter space of supersymmetric models and the nature of physics beyond the Standard Model.

Alternative Models and Beyond

Because there is no direct evidence for supersymmetry, new models and extensions have evolved to address some of the issues and limits that classic supersymmetric theories confront. These alternate models suggest changes to the structure and predictions of supersymmetry, opening up new paths of investigation and potential explanations for present experimental findings.

Split Supersymmetry: Split supersymmetry is an alternate hypothesis in which the superpartners are much heavier than the Standard Model particles. The hierarchy problem, which concerns the huge difference between the electroweak and Planck scales, is handled in this scenario by disconnecting the superpartners from the low-energy spectrum. Split supersymmetry predicts that the superpartners are out the reach of current colliders, which explains the lack of direct detection.

R-Parity Violation: In traditional supersymmetry, the lightest supersymmetric particle's stability is ensured by a conserved quantum number known as R-parity, making it a candidate for dark matter. Alternative models, on the other hand, include cases in which R-parity is violated, allowing for additional decay modes and interactions involving superpartners. These ideas alter the experimental signs of supersymmetry and open up new possibilities for the search for superpartners.

Hidden Sector Supersymmetry: Another alternate model investigates the idea of hidden sector supersymmetry, in which superpartners dwell in a hidden sector that does not directly interact with the visible sector of particles. The hidden sector may have its own gauge symmetries and particle composition, resulting in unique experimental signatures. Hidden sector supersymmetry sheds new light on supersymmetric theories and opens up new paths for investigation.

These different models and extensions alter supersymmetry's predictions and experimental signatures, making the search for superpartners more difficult and necessitating new methods and experimental approaches. They propose many explanations for the lack of direct evidence for supersymmetry at present colliders and encouraged additional research into the nature of physics beyond the Standard Model.

There are also additional theoretical frameworks that go beyond supersymmetry entirely. Extra dimensions, composite models, and string theory are examples of ways to addressing outstanding questions in particle physics. These alternative frameworks open up new avenues for understanding the underlying basis of the cosmos and may shed light on the genesis of mass, dark matter, and other unsolved questions.

To summarise, while supersymmetry research continues to be active, alternative models and extensions offer useful options and possibilities for researching physics beyond the Standard Model. These modifications alter the predictions and experimental signs of supersymmetry and provide potential solutions to the problems that standard supersymmetric theories encounter. Continued theoretical and experimental efforts are required to investigate these alternative models and expand our understanding of fundamental particles and forces.

Future Prospects

As scientists continue to investigate the existence of superpartners and probe deeper into the secrets of the cosmos, the future of supersymmetry research offers immense promise. While existing experimental discoveries may not give solid evidence for supersymmetry, ongoing efforts and future prospects provide exciting opportunities for additional research. We examine some of the future prospects for supersymmetry here:

High-Energy Colliders

Future colliders, such as the High-Luminosity LHC (HL-LHC) and projected linear colliders like the International Linear Collider (ILC) and the Compact Linear Collider (CLIC), are expected to deliver higher collision energy and more luminosity. These sophisticated devices can test a greater variety of mass scales and are more sensitive to the formation of superpartners. They may provide previously unavailable insights on the existence and properties of superpartners.

Indirect Searches

In addition to direct collider searches, indirect approaches play an important role in probing supersymmetry. Astrophysical measurements, such as looking for dark matter annihilation products or cosmic rays, can provide important information about the presence and nature of superpartners. Precision measurements of observables such as rare decays and electric dipole moments can also give indirect evidence for supersymmetry while also shedding information on its parameter space.

Model Building and Phenomenology

Advances in model building and phenomenology continue to be important in the study of supersymmetry. New theoretical frameworks, alternative models, and innovative methods to supersymmetry can offer new insights and predictions for experimental testing. Simulations and data analysis in phenomenological research are critical in refining search tactics and interpreting experimental results.

Cosmological Implications

The impact of supersymmetry for cosmology and the early cosmos are significant. Ongoing studies of the cosmic microwave background radiation, large-scale structure, and other cosmological probes allow for constraints on supersymmetric models. These discoveries can provide crucial insights and limits on the parameter space of supersymmetric theories by analysing the impact of supersymmetry on the evolution of the cosmos.

Synergy with Other Fields

The search for supersymmetry benefits from collaboration between different fields of study. Collaborations and exchanges among particle physics, astrophysics, cosmology, and theoretical physics promote multidisciplinary methods and allow for a more complete knowledge of supersymmetry. The interaction of experiments, observations, and theoretical discoveries improves the search for superpartners and enriches our understanding of nature's fundamental rules.

In summary, while the direct discovery of superpartners has been elusive thus far, the prospects for supersymmetry research in the future remain bright. Advances in high-energy colliders, indirect searches, model construction, and cosmic discoveries provide chances to explore deep into supersymmetric scenarios. The interaction of theory and experiment, as well as collaborations across scientific disciplines, will continue to drive progress in our knowledge of supersymmetry and its role in exposing the secrets of the cosmos.

Supersymmetry and a chance finding of superpartners have significant implications for our knowledge of particle physics, fundamental forces of nature, and cosmic mysteries. While experimental investigations continue, the search for supersymmetry remains an attractive and vibrant area of particle physics research.

5.2 String Theory and Multiverse

Introduction

In this section we begin by discussing string theory, which is a theoretical framework aimed at providing a unified account of all fundamental particles and forces in nature. String theory claims that the fundamental building blocks of the universe are small, vibrating strings rather than point-like particles. We explain the basis for string theory and its potential to reconcile the two cornerstones of modern physics, quantum mechanics and general relativity.

String Vibrations and Extra Dimensions

String theory's fundamental entities are small one-dimensional strings rather than point-like particles. These strings can vibrate in various modes, and the vibrational patterns affect the properties of particles in nature. The string's numerous vibrational modes produce particles with varying masses, charges, and spins. This provides a novel approach to understanding particle variety and interactions.

String theory relies heavily on the concept of extra dimensions. String theory demands the presence of additional spatial dimensions, yet our ordinary experience is limited to three dimensions of space and one dimension of time. These extra dimensions have been compactified or "curled up" in such a way that they are no longer directly visible at our energy scales. The size and form of these extra dimensions are critical for impacting particle characteristics and the forces they encounter.

The concept of extra dimensions has far-reaching implications. It suggests a solution to the hierarchy problem, which is concerned with the sharp contrast in strength between gravity and the other fundamental forces. The seeming weakness of gravity in our observable universe can be explained by spreading out the effects of gravity into these other dimensions.

Furthermore, the concept of extra dimensions gives a foundation for comprehending force unification. At high energies, the forces that appear separate in our everyday experience in higher-dimensional space can be united into a single basic force. This unification provides a look into nature's essential unity and holds the promise of a complete comprehension of the fundamental rules that govern our universe.

String theory's compactification of extra dimensions is a hard and active area of research. The ensuing physics in our observable universe are determined by the exact geometry and size of the extra dimensions. Various compactification strategies, such as Calabi-Yau manifolds, have been suggested, which give a rich mathematical framework for investigating the features of the extra dimensions.

Due to the incredibly high energies necessary to test these scales directly, experimental evidence for the existence of other dimensions and the vibrational modes of strings remains a difficulty. Indirect effects and signatures in high-energy particle colliders and cosmological data, on the other hand, are being actively explored as potential pathways to test string theory predictions and the presence of extra dimensions.

To summarise, string theory requires the existence of extra dimensions and the vibrational modes of strings. The vibrational patterns generate the diverse spectrum of particles, while the extra dimensions provide a framework for unifying forces and answering fundamental physics issues. Exploring the implications of these vibrational modes and extra dimensions is a thriving field of study that has the potential to transform our knowledge of the fundamental nature of the cosmos.

Unified Description and Quantum Gravity

String theory's ability to accommodate gravity inside the framework of quantum physics is one of its most amazing features. Gravity is notoriously difficult to reconcile with quantum mechanics principles in traditional quantum field theories, leading to the long-standing problem of quantum gravity. String theory, on the other hand, provides a unified account of all fundamental forces, including gravity, and thus offers a potential solution to this challenge.

String theory's fundamental entities are microscopic, vibrating strings rather than point-like particles. These strings can move through spacetime, interacting with one another and producing complex patterns. Importantly, gravity emerges naturally inside string theory as one of the fundamental interactions mediated through the interchange of gravitons, which are gravitational excitations.

Unlike other quantum field theories that consider gravity to be an external force acting on quantum particles, string theory considers gravity to be an inherent element of the fabric of spacetime. This means that gravity is incorporated into the same mathematical framework that represents the other interactions, rather than being distinct from them. String theory's outstanding feature is the unification of gravity with the other basic forces.

Gravity's presence in string theory gives a consistent and theoretically well-defined foundation for gravity quantization. This is a big step forward because existing approaches to quantum gravity, such as canonical quantization or loop quantum gravity, have shown to be extremely difficult to quantify gravity. String theory, on the other hand, includes quantum phenomena into gravitational interactions naturally.

String theory seeks to overcome the problem of quantum gravity by treating gravity as a basic force mediated by strings. Attempts to quantify gravity in standard ways find infinities and inconsistencies that make meaningful physical predictions impossible. String theory, on the other hand, has the potential to overcome these issues and produce a consistent quantum theory of gravity.

It is important to highlight that properly comprehending the implications of string theory for quantum gravity, as well as getting experimental evidence to back up its predictions, are ongoing challenges. The energy required to directly probe the string scale or witness the consequences of other dimensions are currently beyond the capability of our technology. String theory, on the other hand, provides a coherent framework that has the potential to deepen our understanding of the underlying nature of the world, including the mysterious domain of quantum gravity.

In summary, string theory gives a unified description of the fundamental forces, including gravity, as well as a potential solution to the quantum gravity problem. It presents a consistent and promising approach to quantizing gravity by treating gravity as an inherent element of the fabric of spacetime and putting it into the mathematical framework of string theory. While experimental evidence is currently sparse, the pursuit of a quantum theory of gravity within the framework of string theory continues to inspire and drive basic theoretical physics research.

String Theory and Multiverse

The multiverse concept arose from the research of string theory and its numerous solutions. String theory models suggest that there may be numerous worlds, each with its own set of physical laws, particle attributes, and fundamental constants. The multiverse is a collection of these worlds that present a variegated landscape of conceivable realities.

Different sections or "pocket universes" within the larger multiversal structure can exhibit diverse traits in the context of the multiverse. For example, the values of fundamental constants such as gravity's strength or the mass of elementary particles may differ from universe to universe. This implies that physics laws and matter and energy attributes can take different forms in different regions of the universe.

The presence of the multiverse raises important issues about our place in the universe and the nature of the reality we observe. The anthropic principle is an intriguing concept relating to the multiverse. According to the anthropic principle, observed aspects of our universe, such as the values of basic constants, are not random, but rather bound by the requirement for the advent of sentient life.

According to the anthropic principle, our world is one of many in the multiverse, and life can only start and evolve in universes with particular physical attributes. In other words, our universe's physical rules and constants appear to be fine-tuned to sustain the presence of sentient beings like ourselves. This concept offers a plausible explanation for why we live in a universe with specified attributes.

The anthropic principle is a tautology, according to critics, because it relies on the existence of observers to make predictions about the features of the universe. Nonetheless, the multiverse and the anthropic principle created exciting scientific debates and conversations regarding the nature of our universe and its place in the greater cosmic landscape.

While the concept of the multiverse comes from specific solutions of string theory, the multiverse itself remains a hypothetical idea, and direct observable evidence remains problematic. Detecting or establishing the existence of other universes within the multiverse poses enormous hurdles because these regions may be beyond our observational grasp due to the restrictions imposed by the speed of light and the expanse of space.

In summary, the multiverse theory proposed by string theory proposes the presence of numerous universes with separate physical rules and attributes. Different parts of the multiverse may have different fundamental constants and particle attributes, creating a colourful tapestry of alternative universes. According to the anthropic principle, our observed universe is one of many in the multiverse that can enable the advent of intelligent life. While the multiverse remains a hypothetical concept, it opens up exciting possibilities and inspires scientific and philosophical inquiry.

Challenges and Current Researches

String theory challenges and ongoing research are focused on several main topics. The search for a unique formulation of string theory that incorporates all physical processes observable in our universe is one of the fundamental problems. String theory currently exists in several variants, including Type I, Type IIA, Type IIB, heterotic, and others, each with its own mathematical framework and set of predictions. The search for an unified framework that includes all of these versions and provides a consistent account of our seen universe is still ongoing.

Another challenge is the investigation of string compactifications and geometries. String theory necessitates the existence of additional spatial dimensions in addition to the known four dimensions of space and time. These extra dimensions must be compacted or coiled up in accordance with observations. The geometry of these compactified dimensions, as well as the distribution of matter and energy inside them, is critical in shaping our universe's physical features and particle spectrum. Exploring the huge landscape of possible compactifications and discovering those that match our universe's observed features is a significant effort.

String theory predictions are also difficult to test experimentally. String theory predicts occurrences at energies well beyond what modern particle colliders can achieve. As a result, direct experimental verification of string theory is currently out of our technological reach. Researchers are investigating indirect testing by investigating the implications of string theory for cosmological phenomena such as cosmic microwave background radiation and galaxy dispersion. These observations shed light on the early universe and may provide proof or restrictions on string theory.

Furthermore, the relationship between string theory and quantum gravity is still being studied. While string theory incorporates gravity into its framework, the specific explanation of how gravity derives from the fundamental string degrees of freedom remains a work in progress. A significant priority remains the development of a full understanding of quantum gravity within the setting of string theory.

In recent years, there has also been a surge of interest in investigating the links between string theory and other branches of physics, such as condensed matter physics and quantum information theory. These interdisciplinary techniques try to find new insights and applications for string theory that go beyond its typical scope.

Finally, the challenges of string theory research include developing a unique formulation that encompasses all observed phenomena, comprehending the geometry of extra dimensions and their role in determining the properties of our universe, and developing experimental tests and indirect probes of string theory predictions. Ongoing theoretical and experimental research efforts continue to push the boundaries of our understanding and shed light on the underlying nature of the cosmos.

Critiques and Alternative Approaches

String theory has faced criticism and other approaches over the years, bringing into question its uniqueness and applicability. One frequent criticism of string theory is the lack of experimental verification or direct empirical data. Because the energy scales needed to actually detect strings are currently unavailable, some contend that string theory is more of a mathematical construct than a testable scientific hypothesis. String theory's lack of empirical validation, critics claim, calls into doubt its validity and predictive potential.

Alternative ways to understanding the nature of quantum gravity have been presented, each with a unique approach to tackling the problem. One such approach is loop quantum gravity, which aims to quantize spacetime by discretizing it into a network of interconnected loops. This method seeks to reconcile general relativity and quantum physics by focusing on geometry quantization. Another alternative technique is causal dynamical triangulation, which discretizes spacetime using triangulations and investigates the statistical behaviour of these triangulations to build a theory of quantum gravity.

These distinct techniques highlight different features of the quantum gravity problem and provide different routes of inquiry. They offer alternate frameworks for addressing fundamental concerns about the nature of space, time, and gravity that are beyond the scope of string theory.

It is critical to recognise that the pursuit of alternative perspectives and critiques is a vital component of scientific investigation. Diverse perspectives and approaches help us understand the cosmos better. While string theory remains the leading option for a theory of everything, it is crucial to find alternative approaches and critically assess different points of view in order to expand our understanding.

Finally, string theory critics point out the lack of direct empirical proof and raise concerns about its testability and uniqueness. Alternative approaches to the problem of quantum gravity, such as loop quantum gravity and causal dynamical triangulation, provide diverse viewpoints on the subject. The scientific community benefits from the diversity of ideas and methodologies because they jointly drive the search for a comprehensive theory capable of unifying all fundamental forces and describing the underlying nature of the cosmos.

In conclusion, string theory represents a remarkable and ambitious effort to uncover the underlying fabric of the universe. Its ability to unite all known forces and reconcile quantum physics with gravity has intrigued scientists and motivated decades of investigation. The concept of strings vibrating in higher dimensions, as well as the formation of the multiverse, have stirred intense debates and encouraged new lines of inquiry.

While there are obstacles to string theory, such as the lack of direct experimental verification and other methodologies that call its uniqueness into doubt, it remains an active and productive area of research. Researchers are still refining the mathematical formalism, experimenting with various string compactifications, and looking for experimental signals that may support or dispute its predictions.

Moreover, string theory has had a far-reaching impact outside of its own discipline, influencing fields such as condensed matter physics and cosmology. String theory concepts and techniques have found unanticipated applications, expanding our understanding of a wide range of phenomena.

String theory research and its implications for the underlying nature of reality exhibit the spirit of scientific research and intellectual curiosity. The effort to solve the universe's secrets necessitates interdisciplinary cooperation, technological developments, and the investigation of alternative theories. As we continue to explore further into the complexities of string theory and its alternatives, we broaden our understanding and open up new avenues for understanding the underlying principles that govern our universe.

Throughout this journey, it is critical to remain open to other points of view, develop debate among scholars, and encourage idea cross-pollination. Whether string theory eventually emerges as the genuine description of the world or leads us to alternative frameworks, the pursuit of understanding the universe at its most fundamental levels is a monument to human curiosity, inventiveness, and the never-ending desire for knowledge.

We are set to discover the mysteries that lay beyond our current comprehension as we stand at the frontier of scientific discovery. String theory continues to inspire and challenge us, leading us to probe the profound nature of our existence and push the boundaries of human understanding. We are getting closer to discovering the ultimate secrets of the universe and exposing the fundamental rules that govern our existence thanks to continual advances and cooperation.

5.3 Quantum Gravity and Black Holes

Introduction

In this section, we will first introduce the theory of quantum gravity, which tries to combine quantum mechanics and general relativity. Quantum gravity aims to characterise the underlying nature of spacetime and gravity at the smallest scales, where quantum mechanics' effects become substantial. We explore the difficulties presented by the incompatibility of these two ideas, as well as the need for a coherent framework that can accommodate both.

Quantum Gravity and the Problem of Singularities

The problem of singularities is central to the challenge of unifying general relativity with quantum physics. When matter collapses under its own gravitational force, it can reach a point of infinite curvature and density known as a singularity, according to general relativity. Singularities are theorised to exist at the centres of black holes, where gravitational forces become infinitely powerful.

The presence of singularities in classical general relativity, on the other hand, presents fundamental issues. In classical physics, infinities usually indicate a theory breakdown, and it is thought that a more thorough description of gravity involving quantum phenomena is required to fix this issue.

Quantum gravity, as a theoretical paradigm, tries to solve the problem of singularities by combining quantum mechanics and general relativity. It is intended that by including quantum effects into the description of gravity, a more full understanding of the nature of singularities, particularly in the context of black holes, can be reached.

The concept of quantum gravitational phenomena at the cores of black holes is one proposed solution to the singularity problem. The application of quantum principles at these extreme settings suggests that the classical concept of a singularity may be replaced by a high energy density zone dominated by quantum gravitational phenomena. These processes may be able to stabilise or alter the structure of the black hole's core, preventing the presence of a real singularity.

Several quantum gravity theoretical approaches, such as loop quantum gravity and string theory, provide insights into the nature of black hole singularities. According to loop quantum gravity, spacetime possesses a discrete structure at the Planck scale, which may prohibit the emergence of singularities. String theory, on the other hand, proposes that the fundamental entities of nature are small vibrating strings rather than point-like particles. The behaviour of these strings towards the centre of a black hole could provide a solution to the singularity problem.

While singularity resolution in the setting of quantum gravity is an active topic of research, it is important to highlight that our current understanding is still limited. The severe circumstances associated with singularities make it difficult to directly probe these locations using conventional experimental techniques. However, physicists continue to investigate the potential implications of quantum gravity on the nature of singularities and the behaviour of black holes using mathematical models, theoretical research, and thought experiments.

Quantum Gravity Approaches

In the effort to bring together quantum mechanics with gravity and build a consistent framework for quantum gravity, various approaches and hypotheses have developed. Now, we will look at some of the most popular approaches:

Loop Quantum Gravity: Loop Quantum Gravity is a classical quantization approach that attempts to quantize the geometry of spacetime itself. It holds that spacetime is made up of discrete, indivisible units or "loops" that connect to form a network. The theory outlines how these loops and their interactions are quantized, creating a distinct and granular description of spacetime. Loop quantum gravity provides insights into the behaviour of matter and gravity in extreme settings and provides a distinct perspective on the nature of space and time at the tiniest scales.

String Theory: String theory claims that the underlying constituents of the cosmos are small, vibrating strings rather than point-like particles. By characterising particles as distinct vibrational modes of these strings, it blends quantum mechanics and general relativity. String theory introduces new spacetime dimensions beyond the traditional three spatial dimensions and one time dimension, with the goal of unifying all known fundamental forces, including gravity. It also anticipates the existence of other particles, such as gravitons, which act as intermediaries between the gravitational pull and the observer. String theory has received a great deal of attention due to its potential to address black hole singularities and give a consistent explanation of quantum gravity.

Causal Dynamical Triangulation (CDT): A non-perturbative approach to quantum gravity is Causal Dynamical Triangulation (CDT). It discretizes spacetime into simplices or triangles and investigates their potential configurations and evolutions. CDT is concerned with the causal structure of spacetime and the preservation of causal linkages between events. CDT proposes to generate a well-defined path integral for quantum gravity by summing over various triangulations and configurations. It sheds light on the quantum behaviour of spacetime as well as the emergence of classical geometry.

Emergent Gravity: According to certain theories, gravity develops from more fundamental principles rather than being a fundamental force. Gravity is viewed as an emergent phenomena emerging from the collective behaviour of other degrees of freedom, such as quantum entanglement or condensed matter systems, in these approaches. Emergent gravity theories attempt to explain macroscopic gravitational behaviour by tracing it back to tiny elements.

It's worth noting that various approaches provide diverse views and mathematical formalisms, and they're still the subject of current research and debate. Each technique introduces its own set of obstacles and potential insights into the quantum gravity problem. Exploring the linkages and potential unification of these ideas is a fascinating topic of study, with the possibility of gaining a more complete understanding of the underlying nature of our universe.

Quantum Information and Black Hole Entropy

The connection between quantum gravity and black hole entropy has given deep insights into the underlying nature of spacetime and the role of information in gravitational systems. Stephen Hawking's fundamental work on black hole radiation and the concept of Hawking radiation was a watershed moment in understanding black hole's quantum behaviour.

Hawking's discovery resulted from his research into the interaction of quantum physics and black holes. By studying the quantum forces surrounding a black hole's event horizon, he established that black holes are not completely black, but emit a feeble radiation known as Hawking radiation. This radiation is caused by quantum fluctuations at the event horizon, where particle-antiparticle pairs are constantly formed and destroyed. Occasionally, one of these particles falls into the black hole while the other escapes to infinity, causing the black hole to lose mass over time.

Most importantly, Hawking's work had far-reaching implications for the concept of black hole entropy. Entropy is a measure of a system's information content that was originally connected with thermodynamic systems. Hawking's simulations, on the other hand, proved that black holes have intrinsic entropy even in the absence of matter or other sources of information. This discovery called into question traditional conceptions of entropy and the fundamental nature of black holes.

The discovery of black hole entropy resulted in the development of the holographic principle, an important idea in comprehending the link between gravity and quantum information. According to the holographic principle, the information content of a region of spacetime can be fully recorded on its boundary, similar to a hologram. In other words, the information contained within a three-dimensional volume can be characterised by degrees of freedom located on the volume's two-dimensional boundary.

This extraordinary concept has far-reaching consequences for our knowledge of spacetime and gravity. It indicates that a lower-dimensional quantum mechanics theory can describe the three-dimensional universe of gravity. The holographic principle has been extensively studied in the context of string theory, and it is closely related to the AdS/CFT correspondence, which establishes a link between theories of gravity in higher-dimensional Anti-de Sitter space and quantum field theories on their boundaries.

The holographic principle implies a close relationship between gravity and quantum information. It indicates that the dynamics of spacetime and gravity arise from the underlying quantum information processing that takes place on its boundary. This has resulted in tremendous advances in understanding the microscopic origins of black hole entropy, as well as increased interest in quantum information theory applied to gravity.

The relationship between quantum gravity, black hole entropy, and the holographic principle is still being researched. It sheds new light on the fundamental nature of spacetime, the nature of information in gravitational systems, and the possibility of unifying quantum physics and gravity. We acquire an improved understanding of the connection between quantum mechanics, information theory, and the mysteries of the gravitational domain by researching black hole entropy and its implications.

Black Hole Information Paradox

The black hole information paradox is an exciting paradox that comes from the interaction of quantum physics and general relativity in the setting of black hole evaporation. Information, according to quantum physics principles, is conserved and cannot be destroyed. The phenomena of Hawking radiation, as explained by Stephen Hawking, suggests that black holes can gradually lose mass and energy through particle emission. This raises the question of what happens to information trapped within an evaporating black hole.

The firewall hypothesis proposes that the region surrounding a black hole's event horizon is highly energetic and forms a "firewall." This is one possible answer to the paradox. This would imply that any infalling observer would be met with an extreme energy barrier that would wipe out any information attempting to enter through. The firewall theory has attracted debate and objections because it appears to defy key laws of general relativity and unitarity, which ensure that information is retained and that observers do not encounter such extreme phenomena near the event horizon.

Another proposed resolution is complementarity, which proposes that various observers can have distinct descriptions of the same physical system based on their reference frames and observations. An observer outside the black hole would see the information stored in the outgoing Hawking radiation, whereas an observer falling into the black hole would see a different account of events. Complementarity permits noticeable discrepancies from various perspectives to coexist, without the need for a physical firewall.

The role of entanglement in the black hole information paradox has also been studied. Entanglement, as mentioned earlier, is a fundamental quantum mechanical phenomena in which two or more particles become coupled to the point that the state of one cannot be represented independently of the state of the other(s). Some researchers believe that the entanglement between the Hawking radiation and the remaining section of the black hole plays an important role in information preservation. This theory proposes that the information encoded in the radiation is somehow connected with the black hole's interior, ensuring that it is not lost.

The resolution of the black hole information paradox is still an active research topic that challenges our knowledge of quantum gravity. It has implications for fundamental physics principles like the nature of spacetime, quantum information behaviour, and the consistency of quantum mechanics with gravity. Researchers are looking into other ways, such as using quantum information theory, applying the holographic principle, and looking for connections between black hole physics and other fields of theoretical physics. The resolution of this problem promises to expand our understanding of fundamental natural laws as well as the nature of information itself.

Experimental Signatures

Because of the extraordinarily high energy scales required, such as the Planck scale, which is approximately 1019 billion electron volts (GeV), direct experimental examination of quantum gravity is difficult. Current particle accelerators, such as the Large Hadron Collider (LHC), operate at energy scales many orders of magnitude less than the Planck scale, making direct detection of quantum gravity effects difficult. However, there are several indirect tests and observational signs that could shed light on the quantum foundation for gravity.

High-energy particle collisions are one way to conduct indirect tests of quantum gravity. Although modern day accelerators achieve energy scales considerably below the Planck scale, certain quantum gravity models, such as string theory, predict the existence of extra particles or signals that could be observed at lower energies. In quantum gravity theories, these comprise imaginary particles called gravitons, which are connected with gravitational force. The discovery of gravitons or other quantum gravity signatures would provide strong evidence for the existence of quantum effects in the gravitational world.

Gravitational wave measurements are another possible method for investigating quantum gravity. Gravitational waves are ripples in spacetime created by huge object acceleration. Gravitational waves from merging black holes and neutron stars have previously been spotted by advanced detectors such as the Laser Interferometer Gravitational-Wave Observatory (LIGO). Scientists seek to acquire insights into the quantum nature of gravity and maybe identify departures from classical general relativity by examining the features and behaviour of gravitational waves.

Cosmological observations also offer a once-in-a-lifetime opportunity to explore quantum gravity phenomena. The early cosmos, especially during the period of cosmic inflation, went through extreme circumstances that are believed to be relevant for quantum gravity phenomena. Observations of cosmic microwave background radiation, large-scale cosmos structure, and primordial gravitational waves can provide vital information about the quantum basis of gravity and perhaps show deviations from classical predictions.

Furthermore, research into black holes holds promise for understanding quantum gravity. Black holes are extreme objects that are considered to have effects from both general relativity and quantum mechanics. Observations of black hole features such as mass, spin, and accretion processes can provide insights into the interaction of gravity and quantum phenomena.

While existing experimental and observational capabilities have limitations in directly investigating quantum gravity, future technological and theoretical advances may open up new areas for research. Building more stronger particle accelerators, such as future colliders, and developing novel detection techniques could lead to energy scales closer to the Planck scale. Furthermore, advances in gravitational wave detectors and cosmic measurements may reveal small traces of quantum gravity.

Finally, while experimental signatures and observational tests of quantum gravity are difficult, they are not impossible. While direct detection of quantum gravity effects remains hard given existing capabilities, indirect tests such as particle collisions, gravitational wave observations, cosmic probes, and black hole studies hold promise for learning more about the quantum basis of gravity. Continued technological improvements, theoretical developments, and interdisciplinary collaborations are required to advance our understanding of quantum gravity and its experimental validation.

Future Directions

The topic of quantum gravity and black holes remains completely unexplored, with many unanswered issues. One of the most difficult difficulties is to create an unified theory that reconciles the laws of quantum physics and general relativity. String theory, loop quantum gravity, and causal dynamical triangulation are some of the ways that have been proposed to address this issue. Finding a unique and complete theory of quantum gravity that incorporates all aspects of black hole physics, on the other hand, is a substantial ongoing effort.

One unanswered question is the nature of spacetime at the Planck scale, where quantum gravitational forces are likely to dominate. At these exceedingly small distances and enormous energy, it is unclear how spacetime itself works. Understanding the microscopic structure and dynamics of spacetime is critical for developing a full theory of quantum gravity. This includes investigating non-commutative geometry, holography, and emergent spacetime from underlying quantum degrees of freedom.

Another critical topic is what happens to information that falls into a black hole. The black hole information paradox raises fundamental questions about information preservation and retrieval in the presence of gravitational collapse. The resolution of this paradox is thought to have profound implications for our knowledge of quantum gravity and the nature of spacetime. Finding a consistent model that reconciles quantum physics, black hole thermodynamics, and unitarity remains a significant difficulty.

Furthermore, the role of quantum entanglement in the context of black holes and quantum gravity is being researched. Understanding the entanglement structure and correlations between degrees of freedom inside and outside the black hole horizon is critical for understanding black holes' quantum nature. The investigation of entanglement entropy and its holographic interpretation via the AdS/CFT correspondence yields exciting results but also raises new challenges.

Interdisciplinary partnerships between physicists, mathematicians, and computer scientists will be required in future directions in the subject of quantum gravity and black holes. This includes creating new mathematical tools and techniques to investigate the intricate properties of quantum gravity, investigating numerical simulations to gain insights into the behaviour of spacetime and black holes, and applying advances in quantum information theory to address quantum entanglement and information challenges.

Technological progress will also be critical in the future exploration of quantum gravity and black holes. More larger particle accelerators, such as the projected Future Circular Collider, might potentially reach energies closer to the Planck scale, providing useful data for testing quantum gravity theories. Advances in gravitational wave detectors, such as space-based observatories, may allow for more precise detection of gravitational waves from sources, providing new insights into the quantum aspects of spacetime.

To summarise, the science of quantum gravity and black holes is a dynamic and quickly evolving area of study. In the future, we hope to learn more about the quantum nature of spacetime, solve the black hole information problem, and investigate the role of entanglement in quantum gravity. The pursuit of a comprehensive theory of quantum gravity necessitates interdisciplinary cooperation, new theoretical frameworks, and technological advances. We hope that by addressing these unanswered problems, we will be able to discover the secrets of the quantum world and acquire fundamental insights into the nature of spacetime and black holes.

5.4 The Holographic Principle

Introduction

The holographic principle and its enormous consequences for our knowledge of spacetime are introduced in this section. According to the holographic principle, the information content of a patch of spacetime can be represented in one less dimension on its boundary. We describe how this concept arose from the study of black hole physics and how it has transformed our understanding of the nature of reality.

Black Holes and Information Paradox

Black holes have long fascinated scientists, not just because of their enormous gravitational attraction, but also because of its significant implications in the context of the holographic principle. The groundbreaking research of Jacob Bekenstein and Stephen Hawking in the 1970s laid the groundwork for our current understanding of black holes as information-carrying entities.

Bekenstein postulated in 1972 that black holes should have an entropy proportionate to their event horizon area. This concept questioned the widely held belief that black holes were just "empty" areas of spacetime with no physical features. Bekenstein's entropy hypothesis proposed that black holes had concealed information about their tiny states.

In 1974, Stephen Hawking made a significant discovery. He demonstrated that black holes were not completely black, but instead emitted thermal radiation, now known as Hawking radiation, which causes them to gradually lose mass and eventually evaporate. This discovery had significant implications because it appeared to imply that information may be lost during the process of black hole evaporation, which seemed to defy the fundamental rules of quantum physics, which state that information must be conserved.

This realisation encouraged heated disputes and gave rise to the black hole information paradox. Information, according to quantum mechanics, cannot be destroyed. However, if a black hole totally evaporates, all of the information it holds appears to be lost with the released radiation. This apparent contradiction between quantum mechanics and general relativity, which controls black hole physics, raises serious concerns regarding the nature of information and its preservation within black holes.

The solution to the black hole information paradox has been the focus of much research. Various approaches and hypotheses have evolved in an attempt to reconcile the seemingly irreconcilable. The holographic concept, which states that information within a black hole is not lost but rather recorded on its perimeter in a lower-dimensional description, is one promising path. This indicates that the physics of a black hole may be fully comprehended by describing its event horizon.

The holographic principle offers a fascinating possibility: information within a black hole is holographically projected onto its perimeter, much like a three-dimensional hologram. This holographic encoding indicates that the entropy of a black hole is governed by the degrees of freedom on its surface rather than by the volume of its interior.

The relationship between black holes and the holographic principle sheds new light on the nature of spacetime and information. It implies that the seemingly separate realms of gravity and quantum mechanics are inextricably linked. Exploring this link is critical not just for resolving the black hole information paradox, but also for furthering our knowledge of the fundamental nature of the universe.

Researchers continue to investigate the holographic principle and its implications in their attempt to uncover the mysteries of black holes and their relationship with information. Scientists want to get a better understanding of how information is encoded and retained within black holes through mathematical analyses, thought experiments, and numerical simulations. The answer to the black hole information paradox has the potential to shed light on the nature of quantum gravity, spacetime's fundamental structure, and the laws that govern our world.

AdS/CFT Correspondence

The Anti-de Sitter/Conformal Field Theory (AdS/CFT) correspondence is one of the most impressive advances in the field of holography. This correspondence, established by Juan Maldacena in 1997, provides a deep relationship between a gravitational theory in a spacetime with negative curvature known as Anti-de Sitter (AdS) space and a conformal field theory (CFT) living on the space's edge.

By determining that a theory of gravity in the majority of AdS space is mathematically identical to a quantum field theory focusing on its boundary, the AdS/CFT correspondence gives a physical realisation of the holographic principle. This duality implies that the dynamics of the associated CFT on the boundary can entirely capture the physics of the gravitational theory in the AdS bulk, and vice versa.

The AdS space, which is distinguished by its negative curvature, is key in this relationship. It provides a curved spacetime that allows for a well-defined gravitational theory with a cosmic constant that is negative. The CFT, on the other hand, is a quantum field theory specified on the AdS space boundary. It is a conformally invariant theory, which means that its physical properties do not change when scaled.

The AdS/CFT correspondence has profound implications and has shown to be an effective tool for investigating strongly interacting quantum systems. When classical theoretical physics and quantum field theory approaches fail, such as in tightly coupled systems, the AdS/CFT correspondence provides a new perspective and computational techniques.

Researchers can acquire insights into previously difficult to analyse occurrences by mapping the gravitational theory in the AdS bulk to the CFT on the boundary. The connection connects the worlds of quantum gravity and quantum field theory, allowing scientists to investigate the gravitational dynamics of strongly interacting systems.

The AdS/CFT correspondence has found applications in a wide range of scientific fields, including high-energy physics and cosmology, as well as condensed matter physics and quantum information theory. It has been used to investigate the behaviour of quark-gluon plasmas, black hole dynamics, the physics of tightly linked electron systems, and even the fundamental nature of spacetime itself.

Ongoing research is helping us understand the AdS/CFT correspondence and its implications. The connection has led to significant advancements in our effort to comprehend the underlying nature of the cosmos and has opened up new areas for theoretical study. It is a strong instrument for investigating the intricate relationships between gravity, quantum field theory, and the holographic nature of spacetime.

Gauge/Gravity Duality

The AdS/CFT correspondence is a subset of a bigger idea known as gauge/gravity duality, which implies a close relationship between quantum field theories (QFTs) and gravity theories. While the AdS/CFT correspondence connects a gravitational theory in AdS space to a CFT on its boundary, gauge/gravity duality has been developed to include a broader range of spacetime geometries and field theories.

The duality between the gravitational theory and the QFT in gauge/gravity duality implies that specific observables and quantities in one theory can be mapped to similar observables in the other theory. This mapping enables researchers to examine tightly coupled quantum systems in terms of classical gravity theories, giving a useful tool for investigating their dynamics.

Gauge/gravity duality has been extended beyond AdS/CFT to various spacetime geometries such as asymptotically de Sitter (dS) space, flat spacetime, and even more exotic backgrounds. These developments have opened up new opportunities for studying holography in a variety of situations, including cosmology, condensed matter physics, and quantum information theory.

The dS/CFT correspondence, for example, offers a holographic duality between a gravity theory in a dS spacetime and a quantum field theory living on its border. This duality provides insights into inflationary physics, the early cosmos, and the cosmological constant problem. It establishes a framework for researching the quantum behaviour of spacetime against a dS background and gives light on the relationship between quantum gravity and the expansion of our universe.

Flat space holography, often known as the "holographic principle for flat spacetime," investigates the potential of a holographic description of quantum field theories in flat spacetime. This expansion seeks to comprehend how holography can be applied to systems without negative curvature, such as those encountered in high-energy particle physics.

The research of gauge/gravity duality in these various circumstances has resulted in fascinating findings and breakthroughs in our knowledge of holography. It has revealed new information about the nature of spacetime, quantum gravity, and the behaviour of tightly coupled systems.

Ongoing study is being conducted to investigate and improve all aspects of gauge/gravity duality. Physicists want to gain a better understanding of the underlying linkages between gravity and quantum field theory by exploring holography in various spacetime geometries and field theory contexts. These advancements have the potential to transform our understanding of the nature of spacetime, information, and fundamental physical laws.

Emergent Spacetime and Quantum Entanglement

The concept of emergent spacetime from quantum entanglement provides an intriguing insight into the nature of spacetime and its relationship to quantum information. The geometry and dynamics of spacetime, according to this theory, emerge as a collective behaviour of underlying quantum degrees of freedom, particularly through the distribution of entanglement.

The holographic principle, as demonstrated by AdS/CFT correspondence and gauge/gravity duality, implies that the information content of a gravitational theory can be encoded on the boundary of its corresponding QFT. This suggests that the fundamental description of spacetime exists on its boundary rather than within its bulk, raising the question of how the bulk spacetime originates from this boundary theory.

Entanglement entropy, a measure of entanglement between quantum system subsystems, is critical in this emerging spacetime picture. It is defined as the von Neumann entropy of a subsystem's reduced density matrix, and it quantifies the degree of entanglement between the subsystem and its complement.

Surprisingly studies have revealed that the entanglement entropy distribution in a quantum system can encode the geometry of the emergent spacetime. A crucial conclusion in the setting of holography is the Ryu-Takayanagi formula, which connects the entanglement entropy of a region in the boundary theory to the size of a comparable minimal surface or a bulk extremal surface in the gravity theory.

This relationship between entanglement entropy and geometry implies that the entanglement structure of a quantum system is crucial in determining the emergent spacetime. It suggests that geometric aspects of spacetime, like as curvature and topology, can be regarded as manifestations of the underlying degrees of freedom's entanglement patterns.

The study of emergent spacetime and quantum entanglement has not only deepened our understanding of spacetime's nature, but has also yielded insights into fundamental features of quantum information theory. It has helped to forge new links between quantum information, quantum gravity, and condensed matter physics, resulting in the creation of innovative ideas and methodologies.

Ongoing research in this field aims to investigate the precise mechanisms by which spacetime emerges from entanglement, to investigate the role of quantum entanglement in various holographic dualities, and to discover new links between entanglement entropy, quantum information, and spacetime geometry. These efforts have the potential to shed light on some of physics' most profound puzzles, such as the nature of quantum gravity, the origin of spacetime, and the fundamental structure of the universe.

Applications in QFT and Condensed Matter Physics

Holography has several applications in quantum field theory and condensed matter physics, providing vital insights into the behaviour of highly coupled systems that are difficult to investigate using typical analytical approaches. Holography provides a powerful computational framework for examining these systems' features by mapping them to gravitational theories in higher-dimensional spacetimes.

The study of quark-gluon plasmas, which are strongly interacting states of matter that match the circumstances encountered in the early universe or in heavy-ion collision studies, is one significant use of holography. Traditional theoretical approaches struggle to characterise the behaviour of such systems, but holography has provided a new perspective, particularly through the AdS/CFT correspondence. It allows researchers to examine the properties of plasma by mapping them to the behaviour of black holes in higher-dimensional spacetime. Holography has provided crucial insights into the thermodynamics, transport characteristics, and phase transitions of quark-gluon plasmas, allowing us to better comprehend these exotic forms of matter.

Holography has also made substantial contributions to the understanding of condensed matter systems. Researchers have got useful insights into the behaviour of tightly coupled electron systems, superconductors, and quantum phase transitions by utilising holographic duality. Holography allows these condensed matter systems to be described in terms of gravitational theories, allowing researchers to investigate their emergent features and collective behaviours. Holography, for example, has provided a foundation for understanding high-temperature superconductivity as well as the mechanism underlying unconventional superconductors.

Furthermore, holography has proven to be a powerful tool for investigating the dynamics of tightly coupled quantum field theories, providing a non-perturbative method. It has provided light on the behaviour of many theories, including tightly coupled gauge theories, conformal field theories, and quantum chromodynamics. Researchers have used holographic methods to determine parameters such as correlation functions, entanglement entropy, and transport coefficients in these theories, yielding new insights into their fundamental features.

Holography's applications in quantum field theory and condensed matter physics continue to grow. Ongoing research intends to investigate unusual holographic dualities, create more precise approaches for researching quantum field theories with holography, and apply holography to a broader spectrum of condensed matter systems. These initiatives have the potential to advance our understanding of strongly correlated systems by opening up new theoretical pathways and potentially leading to practical applications in materials science and technology.

Cosmology and the Holographic Universe

The implications of holography for cosmology and our understanding of the early universe are intriguing. The holographic principle states that all information contained within a patch of spacetime can be encoded on its boundary. In cosmology, this means that knowledge on the cosmos's cosmic horizon may describe the entire history and dynamics of the universe.

The potential for holography to give light on the nature of inflation, a theorised epoch of fast expansion thought to have occurred in the early cosmos, is an important implication in cosmology. The observable uniformity and flatness of the cosmos on vast scales is elegantly explained by inflationary cosmology, but the underlying mechanisms causing inflation remain unknown. According to holography, the dynamics of inflation may be recorded on the cosmic boundary, providing a fresh approach to understanding its origin and the formation of primordial fluctuations that give rise to the universe's large-scale structure.

Furthermore, holography may aid in the resolution of long-standing cosmological mysteries such as the nature of black hole entropy, the information loss paradox, and the singularity problem in the setting of the Big Bang. Researchers have investigated the possibility that the entire evolution of the universe, including the formation and evaporation of black holes, can be understood as an intricate interplay of quantum entanglement and information flow by considering the holographic encoding of information on the cosmological boundary.

In recent years, there has been a surge of interest in applying holography to cosmological models, such as the AdS/CFT correspondence or holographic inflationary models. These methods strive to comprehend cosmic occurrences by connecting them to well-established holographic dualities and quantum information and entropy theories.

Although the application of holography to cosmology is still in its early stages, it has the potential to improve our understanding of the early cosmos, inflation, and the cosmic mysteries that have fascinated physicists for decades. Researchers hope to get new insights into the underlying laws of physics and the nature of spacetime itself by investigating the holographic nature of the cosmos, offering up fascinating opportunities for cosmological exploration.

Challenges and Future Directions

The investigation of the holographic principle is not without difficulties and unanswered problems. While the idea of holography has produced significant insights and links between gravity, quantum field theory, and information theory, the underlying mechanism that gives rise to holography is still being researched.

Understanding the precise process that allows the encoding of the dynamics of a gravitational theory in terms of a boundary quantum field theory is one of the fundamental issues. In some circumstances, the AdS/CFT correspondence has offered a tangible realisation of holography; nevertheless, extending this correspondence to more general spacetime geometries or non-conformal field theories presents significant theoretical challenges.

Another difficulty is using holography to cosmic situations. While there has been progress in understanding holography in the context of inflation and cosmic frontiers, many questions remain unresolved. Further theoretical work and collaboration with observational cosmologists are required to investigate the holographic character of the early universe and its implications for the formation of cosmic structures, the existence of dark energy, and the resolution of cosmological singularities.

Furthermore, understanding the consequences of holography for quantum gravity and the underlying structure of reality is a work in progress. Holography offers a tantalising peek into the profound linkages between quantum mechanics, gravity, and information, but a fully realised and unified framework is still a long way off.

Future directions in the research of the holographic principle will require interdisciplinary collaborations that incorporate ideas from quantum gravity, string theory, cosmology, and quantum information theory. New mathematical tools and techniques, like as tensor networks, entanglement entropy measures, and computational methodologies, will also be critical in expanding our understanding of holography.

Furthermore, holographic experimental and observational testing are critical. Observational evidence for holography in cosmic data, particle physics experiments, or condensed matter systems can provide critical insights into the validity and applicability of holographic principles in various physical regimes.

To summarise, the problems and future objectives in the research of the holographic principle are understanding its core mechanism, expanding it to cosmological settings, and studying its implications for quantum gravity and the fundamental structure of reality. We should expect the holographic principle to play an increasingly crucial role in enhancing our understanding of the fundamental rules regulating the universe as theoretical, observational, and experimental studies continue.

Finally, we emphasise the holographic principle's revolutionary potency in redefining our knowledge of spacetime, information, and the interplay of quantum physics and gravity. The holographic principle provides a comprehensive and unifying framework that has the potential to transform our understanding of the cosmos. We hope to decode the secrets of the holographic nature of spacetime and acquire deeper insights into the fundamental fabric of reality by further studying its implications and pushing the boundaries of our current knowledge.

Conclusion: Embracing Quantum Visions

In this book, we have started on a journey through the fascinating realms of quantum physics, supersymmetry, string theory, quantum gravity, and the holographic principle. We have explored the frontiers of scientific knowledge and explored the mysteries of the quantum world, where our intuitive notions of reality are challenged and new vistas of understanding emerge.

Throughout this exploration, we have encountered profound concepts and revolutionary ideas that have reshaped our understanding of the universe. Quantum mechanics, with its probabilistic nature and wave-particle duality, has revealed a rich tapestry of phenomena that govern the microscopic realm. From the famous double-slit experiment to the enigma of quantum entanglement, we have witnessed the fascinating strangeness of the quantum world.

We have also ventured into the realm of supersymmetry, a theoretical framework that promises to unify the fundamental forces of nature and shed light on the mysterious dark matter. We have explored the intricate web of superpartners, their interactions, and the search for experimental evidence of supersymmetry. While challenges and constraints remain, supersymmetry continues to inspire researchers to push the boundaries of our knowledge.

The exploration of string theory has taken us on a journey through higher-dimensional spacetimes, vibrating strings, and the possibility of a multiverse. We have grappled with the complexities of this mathematical framework and its potential to unify quantum mechanics and gravity. Although string theory is still a subject of ongoing research and debate, its elegance and potential have captivated the imagination of physicists worldwide.

Quantum gravity has been a focal point of our investigation, as we have grappled with the challenge of reconciling quantum mechanics with gravity. The study of black holes, their information paradox, and the quest for a consistent theory of quantum gravity have pushed the boundaries of our understanding. The holographic principle, with its tantalizing connection between quantum field theories and theories of gravity, has provided new insights into the fundamental nature of spacetime and the encoding of information.

In embracing this quantum vision, we have come to appreciate the interconnectedness of different branches of physics and the need for interdisciplinary collaborations. The quest for knowledge requires us to transcend disciplinary boundaries and explore the rich tapestry of ideas and approaches that the quantum world presents.

As we conclude this journey, we recognize that there are still many open questions, challenges, and unexplored territories. The path forward lies in continued research, experimental tests, and the pursuit of new theoretical frameworks. It is through the collective efforts of physicists, mathematicians, and scientists from diverse fields that we will unravel the secrets of the quantum realm and deepen our understanding of the fundamental laws of nature.

Quantum Vision is not just a scientific endeavor; it is an invitation to embrace the beauty and wonder of the universe. It is a call to embrace the unknown, to question our assumptions, and to push the boundaries of our knowledge. It is through this journey that we come to appreciate the intricate tapestry of the cosmos and our place within it.

So let us continue to explore, to question, and to embrace the quantum visions that beckon us towards a deeper understanding of the universe. As we do so, we unlock the potential to transform not only our scientific knowledge but also our perception of reality itself.

Embrace the quantum visions and start on the quest to explore the mysteries that lie at the heart of our existence. The adventure awaits!

Further Reading: Recommended Resources for Further Exploration

The topics covered in this book encompass a vast and ever-evolving field of scientific inquiry. For readers who wish to explore deeper into the subjects discussed, the following list provides a selection of highly regarded books, and resources. These resources offer additional insights, in-depth explanations, and diverse perspectives from renowned experts in their respective fields. Whether you are a student, a researcher, or an enthusiast seeking to expand your knowledge, these resources will serve as valuable references for further exploration.

Quantum Mechanics:

"Principles of Quantum Mechanics" by R. Shankar

"Quantum Mechanics and Path Integrals" by Richard P. Feynman and Albert R. Hibbs

"Quantum Mechanics: Concepts and Applications" by Nouredine Zettili

"Quantum Mechanics: Concepts and Methods" by Asher Peres

"Quantum Mechanics: A Modern Development" by Leslie E. Ballentine

"Quantum Mechanics and Quantum Information" by Michael A. Nielsen and Isaac L. Chuang

Supersymmetry and Particle Physics

"Supersymmetry Demystified" by Patrick Labelle

"Supersymmetry and String Theory: Beyond the Standard Model" by Michael Dine

"Supersymmetry and Supergravity" by Pran Nath

"Gauge/Gravity Duality: Foundations and Applications" by Eric D'Hoker and Daniel Z. Freedman

String Theory and Quantum Gravity

"String Theory and M-Theory: A Modern Introduction" by Katrin Becker, Melanie Becker, and John H. Schwarz

"The Little Book of String Theory" by Steven S. Gubser

"A First Course in String Theory" by Barton Zwiebach

"Three Roads to Quantum Gravity" by Lee Smolin

"The Elegant Universe: Superstrings, Hidden Dimensions, and the Quest for the Ultimate Theory" by Brian Greene

Holography and Quantum Field Theory

"Holographic Quantum Matter" by Sean A. Hartnoll, Andrew Lucas, and Subir Sachdev

"Quantum Field Theory in Condensed Matter Physics" by Alexei M. Tsvelik

"Black Holes, Information, and the String Theory Revolution" by Leonard Susskind and James Lindesay

"The Holographic Universe" by Michael Talbot

"The Black Hole War: My Battle with Stephen Hawking to Make the World Safe for Quantum Mechanics" by Leonard Susskind

These resources provide a solid foundation for exploring the intricacies of quantum mechanics, supersymmetry, string theory, quantum gravity, and the holographic principle. However, please note that the field is continually advancing, and new discoveries and publications emerge regularly. It is recommended to consult scientific journals, arXiv preprints, and attend conferences to stay up to date with the latest research.

Remember, the pursuit of knowledge is a lifelong journey, and the resources provided here are just a starting point. Embrace the excitement of discovery, engage in discussions with experts and fellow enthusiasts, and continue to explore the frontiers of science.

Happy reading and may your exploration of these fascinating topics be filled with new insights and revelations!

Glossary: Key Terms and Concepts

Standard Model: The Standard Model is a theoretical framework in particle physics that describes the fundamental particles and their interactions.

Supersymmetry: Supersymmetry is a theoretical extension of the Standard Model that posits the existence of a new symmetry between bosons and fermions, predicting the existence of superpartners for known particles.

Hierarchy Problem: The hierarchy problem refers to the large discrepancy between the weak scale and the Planck scale, and the fine-tuning required to explain why the weak force is so much weaker than gravity.

Superpartners: Superpartners are hypothetical particles predicted by supersymmetry that have the same quantum numbers as known particles but differ in spin.

Gauge Bosons: Gauge bosons are the force-carrying particles in the Standard Model that mediate the fundamental forces, such as photons for electromagnetism and gluons for the strong force.

Higgs Boson: The Higgs boson is a particle associated with the Higgs field, which gives mass to other particles in the Standard Model.

Collider Experiments: Collider experiments involve accelerating particles to high energies and colliding them together to study their interactions and discover new particles or phenomena.

Naturalness: Naturalness is a concept in particle physics that suggests the values of fundamental parameters should not require excessive fine-tuning or cancellation to explain observed phenomena.

String Theory: String theory is a theoretical framework that postulates that fundamental particles are not point-like but rather tiny vibrating strings. It aims to provide a unified theory of gravity and quantum mechanics.

Multiverse: The multiverse is a hypothetical concept that suggests the existence of multiple universes, each with its own set of physical laws and properties.

Quantum Gravity: Quantum gravity is the field of study that seeks to reconcile quantum mechanics and general relativity, aiming to describe the fundamental nature of spacetime at the smallest scales.

Black Holes: Black holes are regions of spacetime with extremely strong gravitational forces, resulting from the collapse of massive objects. They possess an event horizon beyond which nothing can escape, including light.

Event Horizon: The event horizon is the boundary surrounding a black hole beyond which no information or particles can escape its gravitational pull.

Singularities: Singularities are points of infinite curvature and density that exist at the centers of black holes and are predicted by general relativity.

AdS/CFT Correspondence: The AdS/CFT correspondence is a duality between a gravitational theory in Anti-de Sitter space and a conformal field theory on its boundary. It provides a powerful tool to study strongly interacting quantum systems.

Entanglement: Entanglement is a phenomenon in quantum mechanics where two or more particles become correlated in such a way that the state of one particle cannot be described independently of the others.

Entanglement Entropy: Entanglement entropy is a measure of the amount of entanglement between different parts of a quantum system. It plays a crucial role in the holographic principle and the study of emergent spacetime.

Cosmology: Cosmology is the study of the origin, evolution, and structure of the universe as a whole.

Inflation: Inflation is a period of rapid expansion in the early universe that is believed to explain certain observed features, such as the uniformity of the cosmic microwave background radiation.

Quantum Information: Quantum information is a branch of quantum mechanics that deals with the processing, transmission, and storage of information using quantum systems.